Grade 2

Reveal
MATH®

Differentiation Resource Book

Mc
Graw
Hill

mheducation.com/prek-12

Copyright © 2022 McGraw Hill

Send all inquiries to:
McGraw Hill
8787 Orion Place
Columbus, OH 43240

ISBN: 978-1-26-421062-6
MHID: 1-26-421062-0

Printed in the United States of America.

9 10 11 12 SMN 27 26 25 24 23

Grade 2
Table of Contents

Unit 2
Place Value to 1,000
Lessons

Unit 3
Patterns within Numbers
Lessons

Unit 4
Meanings of Addition and Subtraction

Lessons

Unit 5
Strategies to Fluently Add within 100

Lessons

Unit 6
Strategies to Fluently Subtract within 100

Lessons

Unit 7
Measure and Compare Lengths

Lessons

Unit 8

Measurement: Money and Time

Lessons

Unit 9

Strategies to Add 3-Digit Numbers

Lessons

Unit 10
Strategies to Subtract 3-Digit Numbers

Lessons

Unit 11
Data Analysis

Lessons

Unit 12
Geometric Shapes and Equal Shares

Lessons

Understand Hundreds

Name _____

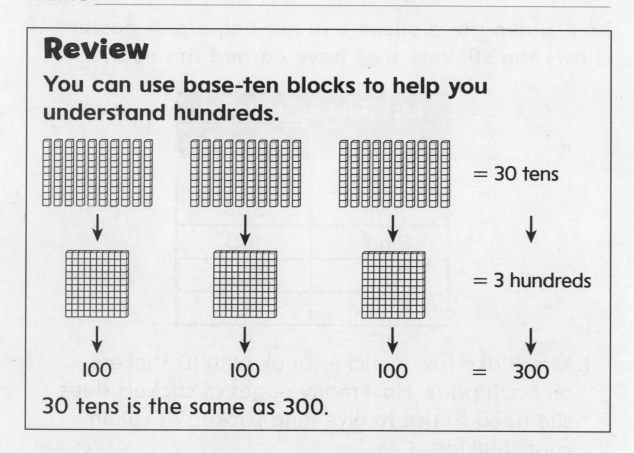

Review

You can use base-ten blocks to help you understand hundreds.

= 30 tens

= 3 hundreds

100　　　100　　　100　　=　300

30 tens is the same as 300.

Match the value with a set of base-ten blocks.

1. 400

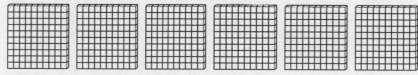

2. 600

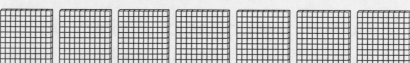

3. 200

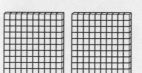

4. 700

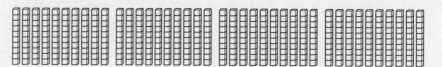

Differentiation Resource Book

1

Understand Hundreds

Name _____

Mrs. Blake gives stickers to her helpers. A poster shows the stickers they have earned this year.

Mrs. Blakes' Helpers	
Name	**Stickers**
André	400
Javier	308
Rina	100
Tai	500
Malik	209

I. Mrs. Blake has a sticker book with 10 stickers on each page. How many pages of stickers does she need to use to give Rina stickers? Explain your thinking.

2. How many pages of stickers does Mrs. Blake need to use to give Malik stickers? Explain your thinking.

Understand 3-Digit Numbers

Name _____

Review

Base-ten blocks and place-value charts can help you describe 3-digit numbers.

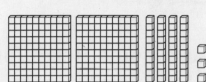

hundreds	tens	ones
2	4	3

243

The number with 2 hundreds, 4 tens, and 3 ones is written as 243.

Write the number shown in two different ways.

1.

hundreds	tens	ones
___	___	___

_____

2.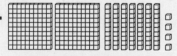

hundreds	tens	ones
___	___	___

_____

3.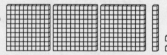

hundreds	tens	ones
___	___	___

_____

4.

hundreds	tens	ones
___	___	___

_____

Understand 3-Digit Numbers

Name _____

1. Lena brings these $100, $10, and $1 bills to the fair.

She buys a ride ticket for $10 and a game ticket for $1. How much money does Lena have left?

$ _____

2. Lena wins these 100 point, 10 point, and 1 point prize tickets.

Lena needs two more 100 point tickets to get a stuffed flamingo. How many prize points does the stuffed flamingo cost?

_____ points

Read and Write Numbers to 1,000

Name _____

Review

You can write numbers in 3 different ways.

hundreds	tens	ones

Expanded form: 200 + 40 + 5

Standard form: 245

Word form: two hundred forty-five

Write the number shown in 3 different ways.

1.

Expanded form: _____ + _____ + _____

Standard form: _____

Word form: _____

2.

Expanded form: _____ + _____ + _____

Standard form: _____

Word form: _____

Read and Write Numbers to 1,000

Name _____

Razi, Kate, and Alana live in the same city. How can you use the clues to decide where each one lives?

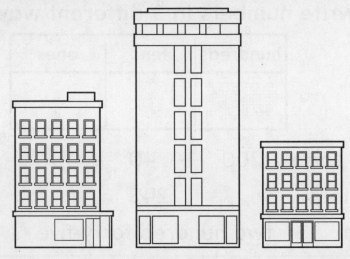

713 Madison Pl.   287 Oak Dr.   574 Lee St.

1. Kate lives in a building with the number 7 in the tens place. Where does Kate live?

2. When Razi writes his building number in word form, part of what he writes is "thirteen." Where does Razi live?

3. Alana likes to tell people her building number is 200 + 80 + 7. Where does Alana live?

Decompose 3-Digit Numbers

Name _____

Review

You can decompose numbers by replacing a base-ten block with other base-ten blocks that have an equal value.

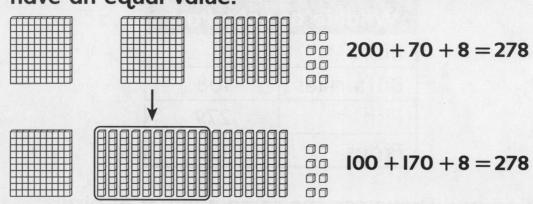

$$200 + 70 + 8 = 278$$

$$100 + 170 + 8 = 278$$

Decompose the number in two different ways.

1. 593

_____ + _____ + _____ = 593

_____ + _____ + _____ = 593

2. 362

_____ + _____ + _____ = 362

_____ + _____ + _____ = 362

3. 745

_____ + _____ + _____ = 745

_____ + _____ + _____ = 745

Decompose 3-Digit Numbers

Name _____

1. Dawn and Vlad are birdwatching at Lake Park. Dawn writes the birds she sees as $300 + 120 + 1$. How many birds does Dawn see? Fill in the number of birds in the table.

Wildlife at Lake Park	
Birds	
Butterflies	108
Fish	279
Frogs	162

2. Vlad and Dawn see the same number of birds. Vlad writes the number of birds, too. It rains and erases his numbers. Now he has ___ $+ 10 +$ ___. What could be his missing numbers?

3. The Lake Park ranger writes down the number of butterflies. He writes $10 + 8$. How can he fix his mistake?

Compare 3-Digit Numbers

Name _____

Review

You can compare 3-digit numbers by looking at each place.

Start with hundreds. If the hundreds are the same, compare tens.

261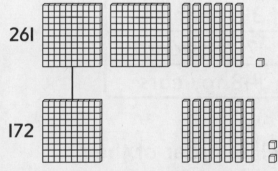

172

261 is greater than 172.

261 > 172

237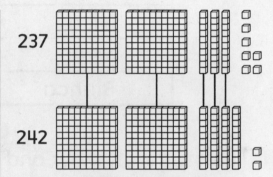

242

237 is less than 242.

237 < 242

Compare the numbers. Fill in <, >, =.

1. 285 ◯ 354

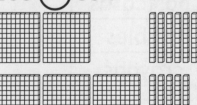

2. 223 ◯ 211

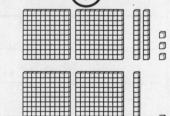

3. 572 ◯ 386

4. 915 ◯ 915

5. 474 ◯ 492

6. 528 ◯ 649

7. 823 ◯ 651

8. 339 ◯ 228

Compare 3-Digit Numbers

Name _____

Mr. An asks his students to bring in collections for show and tell. The table shows the number of objects each student brings.

Grade 2 Student Collections	
Juan	237 crayons
Amy	185 marbles
Nico	315 pennies
Max	179 postcards
Bianca	148 toy cars

1. The students show and tell in order of their number of objects. The person with the most goes last. Write the student names in the order they will show and tell.

Grade 2 Student Collections	
	148 toy cars
	179 postcards
	185 marbles
	237 crayons
	315 pennies

Counting Patterns

Name _____

Review

Number lines can show counting patterns.

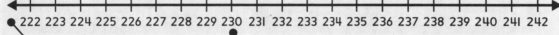

222 223 224 225 226 227 228 229 230 231 232 233 234 235 236 237 238 239 240 241 242

The tens digit stays the same. The ones digit goes up by 1.

Then the tens digit goes up by 1. The ones digit changes to O.

Write the missing numbers.

1.

	242	243	244	245	246	247		249	
251		253	254	255			258	259	260
261	262				266	267	268		270

2.

		773	774	775	776	777		779	780
781	782	783		785	786			789	790
		793	794	795	796	797	798		

What number is missing? Fill in the blank.

3. 507, 508, 509, _____   **4.** 987, 988, 989, _____

5. 397, 398, 399, _____   **6.** 288, 289, 290, _____

Counting Patterns

Name _____

Beau scores the free throw contest, but he is in a hurry and doesn't fill in some numbers.

1. Write the numbers he misses.

Dima	Elise	Elke	Hana	Jean	Wu
	706		993	358	649
413	707		994		
414	708	222			651
	709	223	996		
		224	997	362	653
417	711		998	363	
418				364	
Final Scores					
419		227		365	

2. Beau misses three of Dima's free throws. How many does Dima have? Does she have the fewest?

3. Who makes the most free throws to win? How many free throws does the winner make?

Patterns When Skip Counting by 5s

Name _____

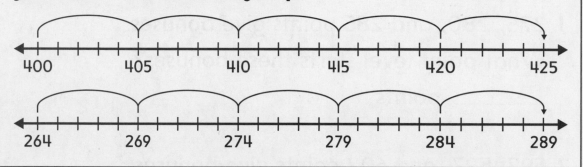

Review

When you skip count by 5 on a number line, you add 5 with each jump.

Skip count by 5s. Fill in the numbers.

1.

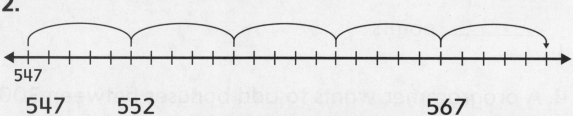

870 _____ 880 _____ _____ 895

2.

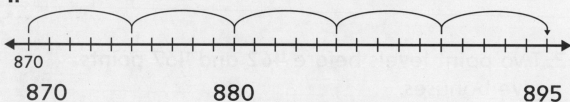

547 552 _____ _____ 567 _____

What number completes the counting pattern?

3. 390, 395, 400, _____   4. 715, 720, 725, _____

5. 176, 181, 186, _____   6. 613, 618, 623, _____

Patterns When Skip Counting by 5s

Name _____

A game gives bonuses when you get to a certain number of points. The game then skip counts by 5 three times to give more bonuses. Fill in the point levels where the bonuses start.

1. 275, 280, and 285 points give bonuses.

What point level starts these bonuses?

_____ points

2. 592, 597, and 602 points give bonuses.

What point level starts these bonuses?

_____ points

3. Two point levels before 462 and 467 points give bonuses.

What point level starts the bonuses?

_____ points

4. A programmer wants to add bonuses between 300 and 399 points. What point levels could he add?

Patterns When Skip Counting by 10s and 100s

Name _____

> ## Review
>
> **You can use a place-value chart to understand skip counting by 10s and 100s.**
>
> When you add 10, only the tens change.
>
> | hundreds | tens | ones |
> |----------|------|------|
> | | | |
>
> 243
>
> | hundreds | tens | ones |
> |----------|------|------|
> | | | |
>
> 253
>
> When you add 100, only the hundreds change.
>
> | hundreds | tens | ones |
> |----------|------|------|
> | | | |
>
> 132
>
> | hundreds | tens | ones |
> |----------|------|------|
> | | | |
>
> 232

Fill in the numbers to complete the number line.

1. Skip count by 10s.

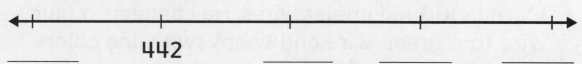

442

2. Skip count by 100s.

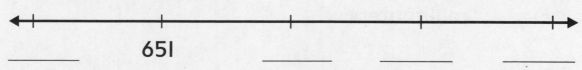

651

Patterns When Skip Counting by 10s and 100s

Name _____

Crispin needs to connect wires for a science project. Help him skip count centimeters by 10s and millimeters by 100s to find the wire lengths.

1. Crispin has a wire 40 centimeters long. He connects 7 more wires. Each new wire is 10 centimeters long. What is the length of the wire now? Explain your thinking.

   _____ centimeters

2. Crispin has another wire wrapped exactly 3 times around two nails that are 100 millimeters apart. What is the length of that wire? Explain your thinking.

   _____ millimeters

3. Crispin has 5 blue 100 millimeter wires and 5 green 100 millimeter wires. He connects a blue wire to a green wire and keeps switching colors until he uses all of the wires. What is the length of the wire now? Explain your thinking.

   _____ millimeters

Understand Even and Odd Numbers

Name _____

Review

You can use a number chart to understand even and odd numbers.

Even numbers of objects can be grouped into pairs with none left over.

You can skip count by 2 to see the pattern in even numbers.

1	2	3	4	5	6	7	8	9	10
11	12	13	14	15	16	17	18	19	20
21	22	23	24	25	26	27	28	29	30
31	32	33	34	35	36	37	38	39	40
41	42	43	44	45	46	47	48	49	50

Circle pairs of objects. Write Even or Odd.

1. _____

2. _____

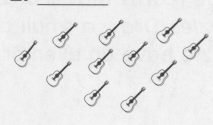

Use counters to show the number.
Circle Even or Odd.

3. 7

   Even   Odd

4. 20

   Even   Odd

5. 12

   Even   Odd

6. 15

   Even   Odd

Understand Even and Odd Numbers

Name _____

A group of students is going on a field trip to the zoo to study the zebras. Answer the questions and circle pairs to show your thinking.

1. Is there an even number or odd number of students going to the zoo?

_____

2. Paige learns that zebras travel in groups called *dazzles*. Draw a small dazzle of zebras. Does your dazzle have an even or odd number of zebras?

Addition Patterns

Name _____

Review

You can use ten-frames to show if a number is even or odd.

The sum in a doubles fact is an even number.

• There are two equal groups with none left over.

The sum in a doubles plus I fact is an odd number.

• There are two equal groups with one left over.

$$4 + 4 = 8 \qquad 3 + 4 = 7$$

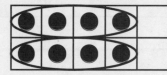   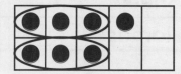

even   odd

Show whether the group is even or odd by circling two equal groups. Write even or odd. Write an addition equation that describes the group.

I. _____   2. _____

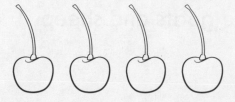   

Show the even number as a sum of a doubles fact.

3. 12 = _____ + _____   4. 6 = _____ + _____

Addition Patterns

Name _____

Jay lives on a farm. The table shows his number of animals. He is buying hay, and a bale of hay feeds 2 animals.

Jay's Farm Animals	
Animal	Number
Cows	9
Goats	5
Horses	9
Sheep	6
Llamas	7

1. How many bales of hay do the horses and cows need? Explain your thinking.

2. How many bales of hay do the goats and sheep need? Explain your thinking.

Patterns with Arrays

Name _____

Review

You can skip count on a number line to find the total number in an array.

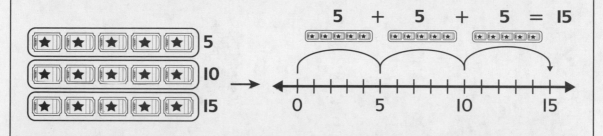

$$5 + 5 + 5 = 15$$

Skip count to find the number of objects in the array.

1.

_____ clovers

2.

_____ party hats

3.

_____ trucks

4.

_____ animals

Patterns with Arrays

Name _____

Sophia is decorating her bedroom. Draw two different arrays she can use to arrange each group of objects.

1. How could Sophia arrange her 8 dolls on shelves?

2. How could she arrange her 20 pictures on a poster?

3. How could Sophia arrange her 16 sports trophies?

Use Arrays to Add

Name _____

Review

You can use arrays to show repeated addition.

Add by rows.

1	2	3
1	2	3
1	2	3
1	2	3

$3 + 3 + 3 + 3 = 12$

Add by columns.

1	1	1
2	2	2
3	3	3
4	4	4

$4 + 4 + 4 = 12$

---

1. Write two equations to describe the array.

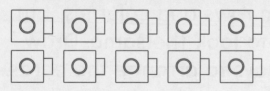

____ + ____ = ____

____ + ____ + ____ + ____ + ____ = ____

2. Shade the array to show 4 rows and 3 columns.
Write two equations to describe it.

_____

_____

Use Arrays to Add

Name _____

Four friends are playing a card game.

1. Carter arranges his cards in 3 rows of 4 cards. Mi arranges her cards in 4 rows of 3 cards. Who has the most cards? Write two equations to show your thinking.

2. Dylan and Isaiah each have 15 cards. They want to arrange their cards in rows and columns, but they do not want them to look the same. Draw to show two different ways to arrange the cards. Write equations to represent each drawing.

Represent and Solve Add To Problems

Name _____

Review

You can use addition to represent and solve a problem in which a number is added to another number.

Jesse has a jar with some marbles. He puts in 5 more marbles. Now there are 32 marbles. How many marbles were in the jar before?

Represent the problem.

$? + 5 = 32$

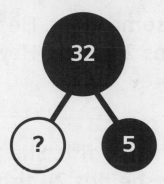

There were **27** marbles in the jar.

I. Fill in the number bond to represent the problem. What equation can represent the problem? Solve.
There are 12 players. Some more players join. Now there are 21 players.

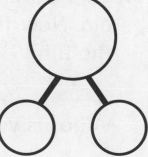

a. Equation: _____

b. Solve: _____

Represent and Solve Add To Problems

Name _____

Solve the problem. Then match the solution with a letter to solve the riddle.

12 = N	15 = E	16 = A	18 = M

1. Mia has 4 toy cars. She gets some more toy cars. Now she has 20 toy cars. How many toy cars does she get?

   _____ _____

2. Theo has some coins. He finds 7 more coins. Now he has 25 coins. How many coins did he have before?

   _____ _____

3. Darla has some fish. She puts 9 more fish in the tank. Now she has 24 fish. How many fish were there before?

   _____ _____

4. There are 6 lemons in a bin. Ty puts more in the bin. Now there are 18 lemons. How many were in the bin?

   _____ _____

   What is yours but your friends use it more?

   _____   _____   _____   _____

     12   16   18   15

Represent and Solve Take From Problems

Name _____

Review

You can use subtraction to represent and solve a problem in which a number is taken from another number.

Jen has 24 books. She gives away some books. Jen has 18 books left. How many books does Jen give away?

Represent the problem.

Part	Part
?	18
Whole	
24	

Write and solve a subtraction equation.

$$24 - ? = 18$$
$$24 - 6 = 18$$

Jen gives away **6** books.

I. Fill in the part-part-whole mat to represent the problem. What subtraction equation can represent the problem? Solve.

Peg makes some party favors. She gives out 9 favors. Now there are 11 left. How many favors does Peg make?

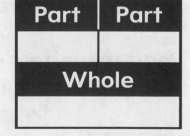

Part	Part
Whole	

a. Equation: _____

b. Solve: _____

Represent and Solve Take From Problems

Name _____

Solve the problem. Write the letter of the alphabet that goes with the number of the answer.
(A = 1, B = 2, C = 3, ...)

1. There are 26 glue sticks. The class throws away 7 old glue sticks. How many glue sticks are there now?   _____ _____

2. There are 34 pieces of paper. Greg uses some. Now there are 18 pieces of paper. How many does Greg use?   _____ _____

3. Li has 20 markers. After class, he finds only 19 markers. How many markers does he lose?   _____ _____

4. Uma buys some paint bottles. She spills 3. Now there are 6 left. How many paint bottles did Uma buy?   _____ _____

5. There are some pens in a box. Fran gets 5 pens out of the box. Now there are 9 pens in the box. How many pens were in the box to start?   _____ _____

Read the letters down the column. In what country was the famous artist Pablo Picasso born?

_____

Solve Two-Step Add To and Take From Problems

Name _____

Review

You can use addition, subtraction, or both to solve a two-step add to or take from problem.

Coach Lyon brings 12 balls to practice. Mae brings 10 balls to practice. Amie brings 4 balls to practice. How many balls are there at practice?

Represent the problem.

tens	ones

Write and solve an addition equation.

$$12 + 10 + 4 = ?$$

$$12 + 10 + 4 = 26$$

There are **26** balls.

I. Draw blocks in the place-value chart to represent the problem. What equation can represent the problem? Solve.

Ari has 24 eggs. He gives 11 to Jade. He gives 9 to Ben. How many does he have now?

tens	ones

a. Equation: _____

b. Solve: _____

Solve Two-Step Add To and Take From Problems

Name _____

The movie theater snack bar tracks food sales.

How can you use the information in the table to solve the problem? Show your thinking.

Item	Sold on Saturday	Sold on Sunday
Pizza Slices	12	5
Water Bottles	40	38
Popcorn Bags	23	19

1. What is the total number of items sold on Saturday?

2. The snack bar had 100 water bottles before the weekend. How many did they have left after the weekend?

3. How many more popcorn bags than pizza slices were sold over the weekend?

Represent and Solve Put Together Problems

Name _____

Review

You can use addition or subtraction to represent and solve problems in which two numbers are put together.

Jack spends 70 minutes mowing. He mows the front yard for 25 minutes and then mows the backyard. How many minutes does he mow the backyard?

Represent the problem.   Write an equation to solve.

70	
25	?

$25 + ? = 70$
$70 - 25 = ?$
$70 - 25 = \mathbf{45}$

Jack spends **45** minutes mowing the backyard.

Write an equation and use a drawing to solve.

1. Keith jogs 45 minutes. He jogs 15 minutes on Saturday and the rest on Sunday. How many minutes does he jog on Sunday?

   a. Equation: _____

   b. Solve: _____

2. The dairy sold 80 gallons of milk. 63 of the gallons were white milk. The rest were chocolate. How many gallons were chocolate?

   a. Equation: _____

   b. Solve: _____

Differentiation Resource Book

Represent and Solve Put Together Problems

Name

A coach recorded walking times for three members of his team. When the members were not walking, they were running.

Name	Minutes Walking	Total Time
Eli	22	53
Gary	10	41
Pam	18	56

Use the information in the table to solve the problem. Show or explain your thinking.

1. How many minutes did Eli spend running?

2. What is the total number of minutes Eli and Pam spent walking?

3. How many more minutes does Pam spend running than Gary?

Represent and Solve Take Apart Problems

Name _____

Review

You can use addition or subtraction to represent and solve a problem in which a total is broken into two groups.

A florist sells 19 roses. 14 are red and the rest are yellow. How many yellow roses are there?

Represent the problem.

Part	Part
14	?
Whole	
19	

Write and solve an equation.

$14 + ? = 19$
$19 - 14 = ?$

$14 + 5 = 19$
$19 - 14 = 5$

There are **5** yellow roses.

I. Fill in the part-part-whole mat to represent the problem. What equation can represent the problem? Solve.

A bowl has 22 pieces of fruit. 18 are bananas. The rest are oranges. How many are oranges?

a. Equation: _____

b. Solve: _____

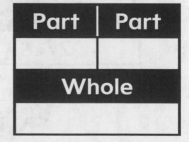

Part	Part
Whole	

Represent and Solve Take Apart Problems

Name _____

What subtraction word problem could you write to match the model? Solve the problem.

1.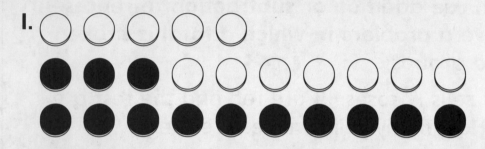

2.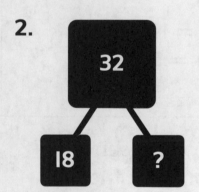

3.

Part	Part
?	12
Whole	
27	

Solve Two-Step Put Together and Take Apart Problems

Name _____

Review

You can use addition, subtraction, or both to solve a two-step add to or take from problem.

The garden has 12 plants. There are some bean plants, 4 tomato plants, and 1 cucumber plant. How many plants are bean plants?

Represent the problem.

12		
4	1	?

Write and solve an equation.

$1 + 4 + ? = 12$ or $12 - 1 - 4 = ?$

$5 + ? = 12$ or $11 - 4 = ?$

$5 + 7 = 12$ or $11 - 4 = 7$

There are **7** bean plants.

1. Draw to represent the problem. What equation can represent the problem? Solve.

   Joe has 14 kites. He gives 3 kites to Malik. He gives 4 kites to Brad. How many kites does he have now?

   a. Equation: _____

   b. Solve: _____

Solve Two-Step Put Together and Take Apart Problems

Name

Today 100 people voted for their favorite new cafeteria foods.

Item	Fourth Graders	Fifth Graders	Teachers
Burrito	12	?	15
Pizza Rolls	3	6	?
Spicy Chicken	?	18	20

How can you use the information in the table to solve the problem?

1. A total of 30 fourth graders voted. How many of them voted for spicy chicken?

2. How many people voted for spicy chicken? Explain your thinking.

3. A total of 10 people voted for pizza rolls. How many teachers voted for pizza rolls?

4. A total of 36 teachers voted. How many fifth graders voted? Explain your thinking.

Lesson 4-7 • Reinforce Understanding

Represent and Solve Compare Problems

Name _____

Review

You can use addition or subtraction to represent and solve a problem in which two numbers are compared.

A bakery has 4 fewer loaves of bread than bran muffins. The bakery has 16 bran muffins. How many loaves of bread are there?

Represent the problem.

m m m m m m m̶ m̶
m m m m m m m̶ m̶
$\underbrace{\hspace{3cm}}$
loaves of bread

There are **12** loaves of bread.

Write and solve an addition or subtraction equation.

$4 + ? = 16$ or $16 − 4 = ?$

$4 + 12 = 16$ or $16 − 4 = 12$

---

1. What equation can represent the problem? Draw to represent the word problem. Solve.

A puppy has 19 toys. A kitten has 12 toys. How many more toys does the puppy have than the kitten?

a. Equation: _____

b. Solve: _____

Represent and Solve Compare Problems

Name _____

Solve the problem. Write the letter that goes with the number of the answer. (A = 1, B = 2, C = 3, ...)

1. A shelf has 45 books and 26 magazines. How many fewer magazines are there than books?

_____ _____

2. An aquarium has 13 neon fish and 8 guppies. How many more neon fish are there than guppies?

_____ _____

3. Sheila has 9 fewer coins than Rick. Rick has 31 coins. How many does Sheila have?

_____ _____

4. Jim has 16 fewer sports cards than Sofia. Sofia has 21 sports cards. How many sports cards does Jim have?

_____ _____

5. A basket has 19 apples and 5 oranges. How many more apples than oranges are there?

_____ _____

Read the letters down the column and write the answer.

What number is even if it has one more letter?

_____

Represent and Solve More Compare Problems

Name _____

Review

You can use addition or subtraction to represent and solve a compare problem.

Kyle eats 6 more grapes than Alex. Kyle eats 15 grapes. How many grapes does Alex eat?

Represent the problem.

Part	Part
6	?
Whole	
15	

Write and solve an addition or subtraction equation.

$6 + ? = 15$ or $15 - 6 = ?$

$6 + 9 = 15$ or $15 - 6 = 9$

Alex eats **9** grapes.

---

1. What equation can represent the problem? Fill in the part-part-whole mat to represent the problem. Then solve.

A drawer has 8 more pairs of socks than T-shirts. The drawer has 20 pairs of socks. How many T-shirts are in the drawer?

a. Equation: _____

Part	Part
Whole	

b. Solve: _____

Represent and Solve More Compare Problems

Name _____

A new pizza shop opens. What two equations can you write to represent the problem? Solve the problem.

1. On Monday, the shop sells 8 more meatball subs than calzones. The shop sells 24 meatball subs. How many calzones does it sell?

2. On Tuesday, the shop makes 20 fewer medium pizzas than large pizzas. The shop makes 68 large pizzas. How many medium pizzas does it make?

3. On Wednesday, the shop makes 9 more salads than pasta dishes. The shop also makes 14 more orders of wings than salads. The shop makes 29 orders of wings. How many salads and pasta dishes does it make?

Represent and Solve Two-Step Problems with Comparison

Name _____

Review

You can represent and solve a two-step problem in which two numbers are compared.

Sam sees 4 more fish than plants. Sam sees 9 plants. How many plants and fish does Sam see?

Represent the problem.

?	
9 plants	9 + 4 fish

Write and solve 2 equations.

number of fish: $9 + 4 = 13$

total number of plants and fish: $9 + 13 = 22$

Sam sees **22** fish and plants.

I. Draw to represent the problem. What equations can represent the problem? Solve.

Sara sees 6 more rocks than shells. She sees 14 shells. How many rocks and shells does she see?

a. Equations: _____

b. Solve: _____

Represent and Solve Two-Step Problems with Comparison

Name _____

A mobile zoo brings some animals to share with students.

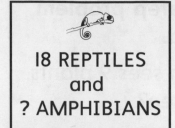 18 REPTILES and ? AMPHIBIANS	 ? INSECTS and 10 BIRDS	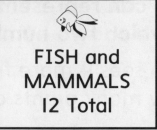 FISH and MAMMALS 12 Total

How can you use the information to show the problem? Show or explain your thinking.

1. There are 7 fewer amphibians than reptiles. How many reptiles and amphibians did the zoo bring?

2. There are 4 more insects than birds. How many insects and birds did the zoo bring?

3. There are 2 more mammals than fish. How many mammals? How many fish?

Solve Two-Step Problems Using Addition and Subtraction

Name _____

Review

You can represent and solve a two-step problem using addition, subtraction, or both.

Ed has 6 songs on his computer. He downloads 9 more songs. He deletes 2 songs. How many songs does Ed have on his computer?

Represent the problem.

Write and solve 2 equations.

downloads: $6 + 9 = 15$

deletes: $15 - 2 = 13$

Ed has **13** songs on his computer.

---

1. Draw to represent the problem. What equations can represent the problem? Solve.

   Mindy has 7 bracelets. She makes 6 more bracelets. Then she gives 2 of bracelets to her sister. How many bracelets does Mindy have now?

   a. Equations: _____

   b. Solve: _____

Solve Two-Step Problems Using Addition and Subtraction

Name _____

Match the word problem with the representation that could help you solve it. Then solve.

1. Ty has 4 bananas. He buys 5 more. Then he eats 2 of them. How many bananas does Ty have now?

    _____

2. Tia has 5 markers. She buys 2 more and then gives 4 of them away. How many markers does Tia have now?

    _____

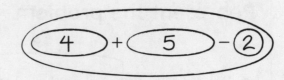

3. Paula has some shells. She finds 4 more shells. Her friend gives her 2 shells. Now she has 11 shells. How many shells does Paula have to begin with?

    _____

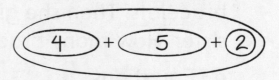

Strategies to Add Fluently within 20

Name _____

Review

You can use strategies to help you add.

$7 + 6 = ?$

Use counters to model each addend.

Make a 10 and count on to find the sum.

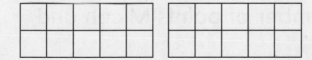

$7 + 6 = 13$

What is the sum? Use the ten-frames to solve.

1. $6 + 9 =$ _____

2. $9 + 7 =$ _____

 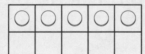

What is the sum?

3. $5 + 7 =$ _____   4. $8 + 9 =$ _____

Strategies to Add Fluently within 20

Name _____

Micah and Finn play on the same basketball team. How can you use the information in the table and a strategy to find each sum? Explain your thinking.

Player	Micah	Finn
Rebounds	2	13
Free Throws	4	7
Points	8	9

1. How many free throws did Micah and Finn make?

_____

2. How many rebounds did Micah and Finn get?

_____

3. What is the total number of points Micah and Finn scored?

_____

More Strategies to Add Fluently within 20

Name _____

Review

Doubles facts can help you find the sum of near doubles facts.

$7 + 8 = ?$

● ● ● ● ● ● ●
● ● ● ● ● ● ● ○

You know $7 + 7 = 14$.

$7 + 8$ is 1 more than $7 + 7$.

So, $7 + 8 = $ **15**.

$7 + 9 = ?$

● ● ● ● ● ● ●
● ● ● ● ● ● ● ○ ○

You know $7 + 7 = 14$.

$7 + 9$ is 2 more than $7 + 7$.

So, $7 + 9 = $ **16**.

How can you use doubles facts to find the sum?

1. $3 + 5 = ?$

● ● ●
● ● ● ○ ○

$3 + 3 = $ _____

$3 + 5$ is _____ more than $3 + 3$.

So, $3 + 5 = $ _____.

2. $6 + 7 = ?$

● ● ● ● ● ●
● ● ● ● ● ● ○

$6 + 6 = $ _____

$6 + 7$ is _____ more than $6 + 6$.

So, $6 + 7 = $ _____.

What is the sum?

3. $6 + 8 = $ _____

4. $8 + 9 = $ _____

More Strategies to Add Fluently within 20

Name _____

Eli works at a grocery store. How can you use a doubles fact to find the sum? Show your thinking.

1. Eli unpacks each of these boxes. How many boxes does he unpack?

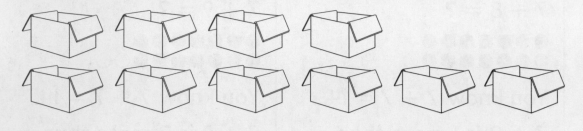

_____

2. There are 8 quarts of milk in the refrigerator. Eli puts 10 more quarts of milk in the refrigerator. How many quarts are in the refrigerator now?

_____

Represent Addition with 2-Digit Numbers

Name _____

Review

Base-ten blocks can help you add.

Add the ones. Then add the tens.

Regroup 10 ones as 1 ten, if needed.

$$44 + 37 = ?$$

Tens	Ones
7 tens	11 ones

$$44 + 37 = 81$$

Tens	Ones
8 tens	1 one

What is the sum? Draw to show your thinking.

1. $32 + 64 =$ _____

tens	ones

2. $41 + 59 =$ _____

tens	ones

What is the sum?

3. $63 + 24 =$ _____

4. $55 + 18 =$ _____

Represent Addition with 2-Digit Numbers

Name _____

Annik is building a city using building blocks. The table is missing the ones digit for the orange blocks. How can you make a drawing to help solve each problem? Explain your thinking.

Color	Number
Blue	23
Green	35
Orange	3_
Purple	13
Red	29
Yellow	41

1. How many blue, red, and purple blocks does Annik use to build his city?

2. Annik regroups to add the number of orange and yellow blocks. How many orange blocks are there? How many orange and yellow blocks are there?

Use Properties to Add

Name _____

Review

Addends can be added in any order. The sum is the same.

$17 + 22 = 39$

$22 + 17 = 39$

What is the sum?

1. $33 + 12 =$ _____

$12 + 33 =$ _____

2. $16 + 24 =$ _____

$24 + 16 =$ _____

3. $53 + 25 =$ _____

$25 + 53 =$ _____

4. $35 + 46 =$ _____

$46 + 35 =$ _____

Use Properties to Add

Name _____

**Match the sets of base-ten blocks with the same
sum. Draw the missing set of base-ten blocks.**

1.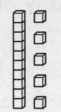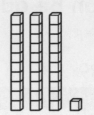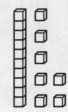

2.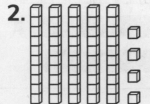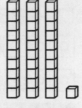

3.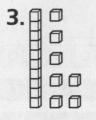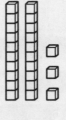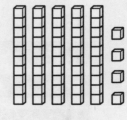

4.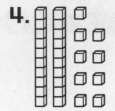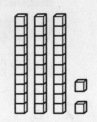

5. Write two equations about one of the matching
   sets of base-ten blocks. How do the base-ten
   blocks show addends can be added in any order?

Decompose Two Addends to Add

Name _____

Review

Decomposing two addends by place value to find partial sums can help you add 2-digit numbers.

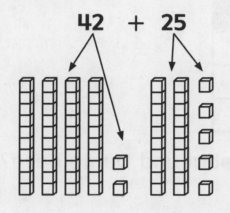

42 + 25

Add the tens.
40 + 20 = 60

Add the ones.
2 + 5 = 7

Add the partial sums.
60 + 7 = 67

I. How can you decompose both addends by place value? Draw base-ten blocks and find the sum.

54 + 38 = ?

Add the tens: _____ + _____ = _____

Add the ones: _____ + _____ = _____

Add the partial sums: _____ + _____ = _____

Decompose Two Addends to Add

Name _____

Three friends are making necklaces. They each have some blue, green, and red beads. How can you use the information from the table to decompose two addends and find the sum? Explain your thinking.

	Blue	Green	Red
Amdal	26	19	13
Gwen	31	15	48
Dante	22	27	38

1. How many green beads do Amdal and Dante have?

_____

2. How many blue beads do Gwen and Dante have?

_____

3. How many red beads do Gwen and Amdal have?

_____

Use a Number Line to Add

Name _____

Review

You can use a number line to add.

The cubes below the number line show the addends.

11 + 7 = ?

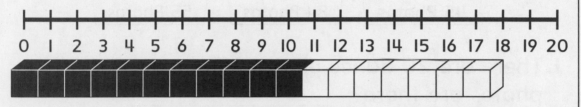

11 + 7 = **18**

1. What is the sum? Use the cubes and number line.

13 + 6 = _____

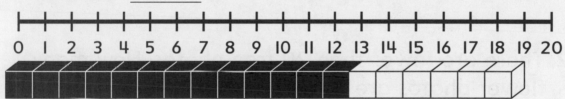

What is the sum? Use the bars and number line.

2. 9 + 8 = _____

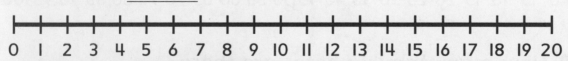

| 9 | 8 |

3. 22 + 27 = _____

| 22 | 27 |

Use a Number Line to Add

Name _____

Jaya is sorting her photos into bins. She can use the number lines to find the number of photos in each bin. Use the number line to help her find the sum.

FLOWERS and TREES 45 Photos   FRIENDS and FAMILY 84 Photos   DANCE and TUMBLING 55 Photos

I. There are 20 dance photos. How many tumbling photos are there? _____

| 35 | 20 |

2. There are fewer than 20 tree photos. How many flower photos are there? _____

3. How many family photos are there?

_____ or _____

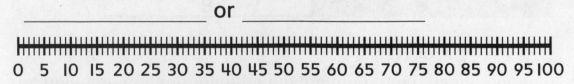

| 48 | 36 |

Decompose One Addend to Add

Name _____

Review

Adjusting addends can make them easier to add.

$42 + 37 = ?$

Decompose 37 into tens and ones.　　　$37 = 30 + 7$

　　　　　　　　　Add the tens.　　$42 + 30 = 72$

　　　　　　　　　Add the ones.　　$72 + 7 = 79$

$42 + 37 = \mathbf{79}$

How can you decompose one addend to help you find the sum? Write the sum.

1. $38 + 44 =$ _____

   $44 =$ _____ $+$ _____

2. $57 + 18 =$ _____

   $18 =$ _____ $+$ _____

3. $59 + 24 =$ _____

   $24 =$ _____ $+$ _____

4. $26 + 28 =$ _____

   $28 =$ _____ $+$ _____

Decompose One Addend to Add

Name _____

Argen and Selena go to the store, where they buy apples for a school party. They need 60 apples for the party. Help them find out if they have enough.

I.a. Argen gets 23 green apples. Selena gets 44 red apples. How many apples did they get? How can you draw a number line to help you decompose one addend to solve?

b. How can you check your answer by drawing a number line and decomposing the second addend?

c. Do Argen and Selena have enough apples for the party? Explain your thinking.

Adjust Addends to Add

Name _____

Review

You can adjust addends to make friendly numbers, which have 0 and 5 in the ones place.

17 + 28 = ?	17 + 28 = ?
Move 3 from 28 to 17.	Move 2 from 17 to 28.

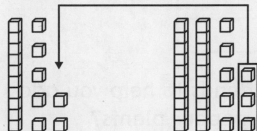 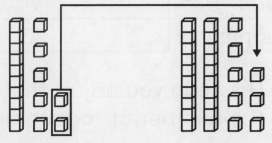

10 ones is 1 ten. 10 ones is 1 ten.

 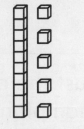

20 + 25 = 45	15 + 30 = 45
17 + 28 = **45**	17 + 28 = **45**

How can you draw base-ten blocks to help you add?

1. 29 + 36   2. 43 + 47

_____ + _____ = _____   _____ + _____ = _____

Adjust Addends to Add

Name

A farmer is planting his garden. How can you adjust the addends to help you find the sum? Explain.

Plant	Number of Plants
Tomato	38
Cucumber	26
Bean	7
Pepper	23

1. How can you adjust the addends to help you add the number of tomato and pepper plants?

2. How can you adjust the addends to help you add the number of bean plants and cucumber plants?

3. Write another addition problem using 2 of the types of plants. How you could adjust the addends to solve the problem?

Add More Than Two Numbers

Name _____

Review

One way to add more than two 2-digit numbers is to decompose the addends.

$$22 \quad + \quad 35 \quad + \quad 19 = ?$$

(20) + [2]   (30) + [5]   (10) + [9]

The tens are circled.
Add the tens. $20 + 30 + 10 = 60$

The ones are in a box.
Add the ones. $2 + 5 + 9 = 16$

Add the partial sums. $60 + 16 = 76$

So, $22 + 35 + 19 = \mathbf{76}$

How can you decompose the addends and add the partial sums?

1. $41 + 39 + 18 = ?$	**2.** $26 + 34 + 17 = ?$
Add the tens.	Add the tens.
_____	_____
Add the ones.	Add the ones.
_____	_____
Add the partial sums.	Add the partial sums.
_____	_____
$41 + 39 + 18 =$ ____	$26 + 34 + 17 =$ ____

Add More Than Two Numbers

Name

Ari works at Madison Mini Golf. How can you write an equation and adjust addends to find the sum?

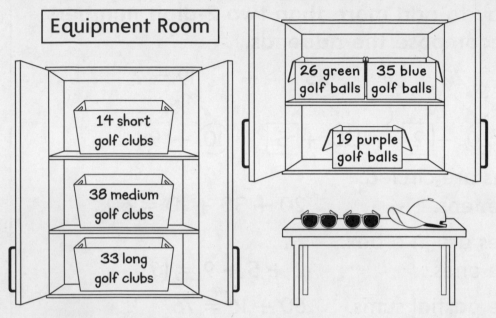

Equipment Room

14 short golf clubs

38 medium golf clubs

33 long golf clubs

26 green golf balls

35 blue golf balls

19 purple golf balls

1. Ari has 26 green, 35 blue, and 19 purple golf balls to sort by color. How many golf balls are there?

2. Ari sorts 14 short, 38 medium, and 33 long golf clubs by length. How many golf clubs are there?

Solving One- and Two-Step Problems Using Addition

Name _____

Review

You can use addition strategies to solve one- and two-step word problems.

Keisha is reading. On Monday she reads 15 pages. On Tuesday she reads 26 pages. On Friday she reads 21 pages. How many pages does she read?

Add the number of pages Keisha reads each day. You can decompose each addend to add.

$$15 \quad + \quad 26 \quad + \quad 21$$
$$\underline{10} + 5 \quad + \quad \underline{20} + 6 \quad + \quad \underline{20} + 1$$

Add the tens. $\underline{10} + \underline{20} + \underline{20} = \underline{50}$

Add the ones. $5 + 6 + 1 = 12$

Then find the sum. $\underline{50} + 12 = 62$

Keisha reads 62 pages.

I. How can you use a strategy to solve the problem?

Paulo runs for 22 minutes. Greta runs for 36 minutes. Enid runs for 27 minutes. How many total minutes do they run?

Solving One- and Two-Step Problems Using Addition

Name _____

Devon, Velia, and Nico are collecting boxes of dog and cat treats for the animal shelter. How can you use the information in the table and an addition strategy to solve the problem? Explain your thinking.

Animal Shelter Collection	Dog Treats	Cat Treats
Devon	23	14
Velia	34	16
Nico	19	25

1. How many boxes of dog treats were collected in all?

   _____

2. How many boxes of cat treats were collected in all?

   _____

3. How can you check the total number of boxes of dog treats collected using a different strategy?

Strategies to Subtract Fluently within 20

Name _____

Review

You can count back with counters to subtract within 20.

$$12 - 4 = ?$$

Use counters to show the total.

●●●●●●●● ⊗⊗⊗⊗
                     8 9 10 11 12

Then cross out and count back to find the difference.

$$12 - 4 = \mathbf{8}$$

How can you use a strategy to subtract? Show your work. Fill in the difference.

1. $11 - 5 =$ _____

             6   7   8   9 10 11

2. $13 - 8 =$ _____

What is the difference?

3. $14 - 5 =$ _____   4. $19 - 6 =$ _____

Strategies to Subtract Fluently within 20

Name _____

Ellen keeps her sports card collections in 3 different books. How can you use a subtraction strategy to help you solve each problem? Explain your thinking.

Baseball Cards   **Hockey Cards**   **Soccer Cards**

 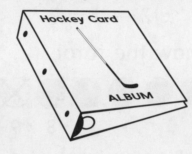 

1. Ellen has 11 hockey and baseball cards. Write the number of hockey cards.

   _____

2. Ellen has 20 sports cards. Write the number of soccer cards.

   _____

3. How many more cards does Ellen need for 2 full pages of hockey cards?

   _____

More Strategies to Subtract Fluently within 20

Name _____

Review

You can use addition doubles facts and addition near doubles facts to subtract.

$$12 - 5$$

You know $6 + 6 = 12$. So, $12 - 6 = 6$.
You need to find $12 - 5$. 5 is one less than 6.
So, $12 - 5 = 6 + 1$, or $12 - 5 = 7$.

How can you use doubles and near doubles facts to find the difference?

1. $15 - 7$

$14 - 7 =$ _____

15 is _____ more than 14.

So, $15 - 7 =$ _____.

2. $10 - 4$

$10 - 5 =$ _____

5 is _____ more than 4.

So, $10 - 4 =$ _____.

What is the difference?

3. $9 - 4 =$ _____   **4.** $17 - 8 =$ _____

More Strategies to Subtract Fluently within 20

Name _____

Rae sells jewelry. How can you use subtraction strategies to find the difference?

1. Rae has 11 rings. Rae sells 4 rings. Show the difference and explain your thinking.

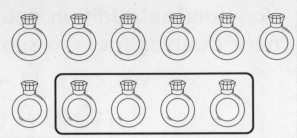

2. Rae has 15 necklaces. She sells 9 necklaces. Show the difference and explain your thinking.

3. Write a word problem about Rae selling her jewelry. Explain how to solve the problem.

Represent Subtraction with 2-Digit Numbers

Name _____

Review

You can represent 2-digit subtraction situations.

Subtract the ones. Then subtract the tens.

$$35 - 21 = ?$$

tens	ones

$$35 - 21 = 14$$

How can you represent the 2-digit subtraction situation? Fill in the difference.

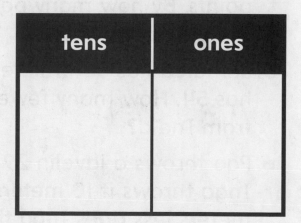

tens	ones

1. $68 - 27 =$ _____

What is the difference?

2. $88 - 54 =$ _____   3. $27 - 16 =$ _____

Represent Subtraction with 2-Digit Numbers

Name _____

Solve the problem. Write the letter that goes with the number of the answer. (A = I, B = 2, C = 3, ...)

1. Ty throws a shot put 23 meters. Fran throws it 22 meters. What is the difference in lengths? _____ _____

2. The distance for Olympic archery is 77 yards. An indoor distance is 57 yards less. How many yards is the indoor distance? _____ _____

3. Egypt won 19 medals. Japan won 11 medals. How many more medals did Egypt win than Japan? _____ _____

4. The U.S. team scored 99 points. The Canadian team scored 94 points. By how many points did the U.S. team win? _____ _____

5. Jamaica has 68 athletes. India has 54. How many fewer are from India? _____ _____

6. Pao throws a javelin 29 meters. Inga throws it 10 meters. How many meters less does Inga throw it? _____ _____

Read the letters down the column. What city held the first Olympics? _____

Represent 2-Digit Subtraction with Regrouping

Name _____

Review

You can use base-ten shorthand to represent subtraction.

$42 - 26 = ?$	Regroup I ten.	Take away 2 tens and 6 ones.
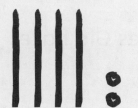		 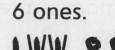 $42 - 26 = 16$

What is the difference? Draw base-ten shorthand.

1. $53 - 37 =$ _____   2. $34 - 15 =$ _____

What is the difference?

3. $63 - 27 =$ _____   4. $50 - 22 =$ _____

Represent 2-Digit Subtraction with Regrouping

Name _____

Marty, Lily, and Gio all collect postcards. How can you use the information in the table to subtract and solve the problem? Explain your thinking.

	Wildlife	City
Gio	41	38
Lily	59	25
Marty	28	

1. How many more wildlife postcards does Gio have than Marty? _____ postcards

2. How many more wildlife postcards does Lily have than city postcards? _____ postcards

3. In all, Lily and Marty have 37 city postcards. Does Marty have more wildlife postcards or city postcards? How many more?

   Marty has _____ more _____ postcards.

Use a Number Line to Subtract

Name _____

Review

You can use a number line to subtract.

The cubes show the whole and I part of the subtraction equation.

$$11 - 4 = ?$$

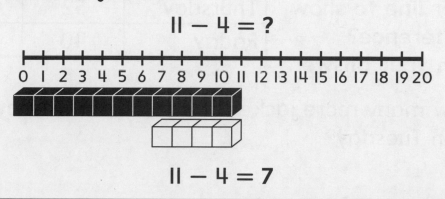

$$11 - 4 = 7$$

I. What is the difference? Look at the cubes to help.

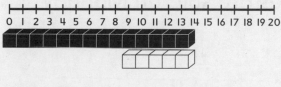

$$14 - 5 = \underline{\qquad}$$

2. What is the difference? Look at the bars to help.

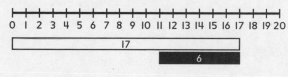

$$17 - 6 = \underline{\qquad}$$

Use a Number Line to Subtract

Name _____

Malena sells jackets and gloves at a store. How can you use the information in the table and draw a number line to show the difference? Explain your answer.

Day	Jackets	Gloves
Monday	34	15
Tuesday	19	24
Wednesday	48	31
Thursday	52	23
Friday	40	33

1. How many more jackets are sold on Thursday than Tuesday?

2. How many fewer gloves are sold on Monday than Friday?

3. How many more jackets than gloves are sold on Wednesday?

Decompose Numbers to Subtract

Name _____

Review

You can decompose a number to subtract.

$$52 - 36 = ?$$

Subtract 2 from 52 to make 50, a friendly number.

Then decompose 36 to include 2. $36 = 2 + 30 + 4$

Make jumps on a number line to subtract

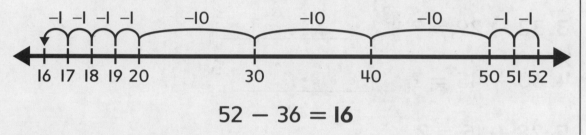

$$52 - 36 = 16$$

---

1. How can you decompose to subtract? Show your work on a number line.

   $49 - 22 = ?$ _____

   $22 =$ _____ + _____ + _____

How can you decompose to subtract?

2. $39 - 16 = ?$ _____

   $16 =$ _____ + _____ + _____

3. $76 - 48 = ?$ _____

   $48 =$ _____ + _____ + _____

Decompose Numbers to Subtract

Name _____

How can you decompose by place value to subtract? Write the letter that goes with the number of the answer.
(A = 1, B = 2, C = 3, ...)

1. 51 − 47 = ?   _____ _____

2. 78 − 73 = ?   _____ _____

3. 32 − 29 = ?   _____ _____

4. 60 − 45 = ?   _____ _____

5. 28 − 15 = ?   _____ _____

6. 45 − 29 = ?   _____ _____

7. 33 − 18 = ?   _____ _____

8. 95 − 76 = ?   _____ _____

9. 22 − 17 = ?   _____ _____

Read the letters down the column. What word means
to break down into parts?

_____

Adjust Numbers to Subtract

Name _____

Review

You can adjust addends to make them friendlier to subtract.

34 − 18 = ?

Subtract 4 from each number.	Decompose I ten as 10 ones.

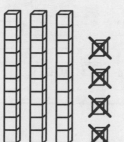 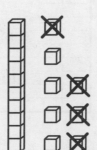 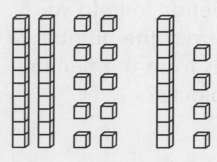

34 − 18 = 16

How can you draw base-ten shorthand to show how to adjust the numbers to make friendly numbers?

I. 43 − 22

2. 29 − 14

___ − ___ = ___   ___ − ___ = ___

Adjust Numbers to Subtract

Name

A toymaker is planning what to make for next season. Explain the strategy you used to find the difference.

16 dolls   34 blocks

47 planes   23 boats

1. How can you show one way to adjust the addends to help you subtract the number of dolls from the number of blocks?

2. How can you show one way to adjust the addends to help you subtract the number of boats from the number of planes?

3. How can you write another subtraction problem using 2 of the types of toys? Explain how you could adjust the addends to solve the problem.

Relate Addition to Subtraction

Name _____

Review

You can solve a subtraction equation by writing it as an addition equation with an unknown addend.

$$41 - 27 = ? \qquad 27 + ? = 41$$

Use a number line. Count the jumps.

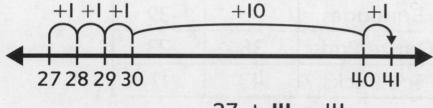

$$27 + 14 = 41$$

What related addition equation can you use to find the difference? Fill in the equation.

1. $65 - 39 = ?$ _____ + _____ = _____

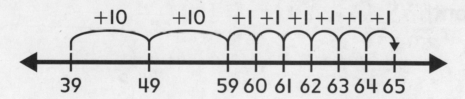

2. $72 - 35 =$ _____

Addition: _____

3. $38 - 16 =$ _____

Addition: _____

Relate Addition to Subtraction

Name _____

The school nurse is counting supplies at the end of the month. How can you use the information from the table to write a subtraction equation? Write a related addition equation to solve. Find the unknown.

Supply	Start	End
Bandages	75	39
Gauze Pads	38	23
Ice Packs	43	17

1. How many bandages were used during the month?

2. How many gauze pads were used during the month?

3. How many more bandages than ice packs were used during the month?

Solve One-Step Problems Using Subtraction

Name _____

Review

You can use subtraction strategies to solve one-step word problems.

Mel donates 46 books. She gives 27 books to a hospital and the rest to a library. How many books does she donate to the library?

Represent the total.

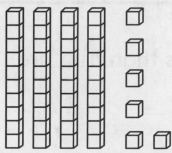

Cross out blocks.

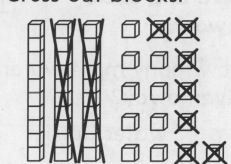

Mel donates 19 books to the library.

How can you use a subtraction strategy to solve the word problem? Explain your thinking.

1. Rio runs for 36 minutes. Em runs for 19 minutes. For how many more minutes does Rio run?

2. Edith counts 31 daisies. Marcus counts 12 roses. How many more flowers does Edith count?

Solve One-Step Problems Using Subtraction

Name _____

A company awards gifts at the end of a 5K race. How can you use the information from the table to solve the word problems? Use any subtraction strategy to solve and show your work.

Product	Number Awarded
Cell Phone Pouches	44
Fruits	68
Granola Bars	76
Water Bottles	95

1. How many more water bottles than fruit are given away?

    _____ water bottles

2. Choose two other products to compare and show the difference.

Solve Two-Step Problems Using Subtraction

Name _____

Review

You can solve two-step subtraction word problems with strategies you know.

Brent has 40 pencils. He gives 13 away. He loses 6 pencils. How many pencils does Brent have left? Decompose an addend by place value.

$40 - 13 - 6$

$10 + 3$ Count back the tens and ones from 40.

$40 - 10 - 3 - 6 = ?$

$30 - 3 - 6 = ?$

$27 - 6 = 21$

Brent has 21 pencils left.

Solve the problem. Explain your thinking with a subtraction strategy.

1. There are 92 balls in a bin. Jules takes 43 out. David takes 35 out. How many balls are in the bin?

2. A florist has 72 flowers. 28 flowers are roses and 17 are carnations. The rest are lilies. How many are lilies?

Solve Two-Step Problems Using Subtraction

Name

A company delivers 97 meals to families in 5 cities each week. How can you use the information from the table and subtraction to represent and solve the problem? Explain your thinking.

City	Number of Meals
Atmore	26
Clayton	15
Jasper	21
Marion	12
Selma	23

1. Suppose deliveries go to Atmore and Clayton first. How many meals still need to be delivered?

2. Suppose deliveries go to Jasper and Marion first. How many meals still need to be delivered?

3. Choose 2 other starting cities. How many meals will be left to deliver after those cities are delivered to?

Measure Length with Inches

Name _____

Review

You can use paper clips to measure length in inches. A paper clip is 1 inch long.

What is the length of the peapod?

Place paper clips end-to-end under the peapod. Count the paper clips.

The peapod is 4 inches long.

What is the length of the object in inches?

1.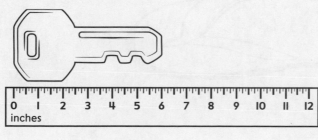

_____ inches

2.

_____ inches

3.

_____ inches

Measure Length with Inches

Name _____

Whose shoe is it? Use an inch ruler to measure.

Kathy	Lucy	Brent
4 inches	5 inches	3 inches

1.

_____

2.

_____

3.

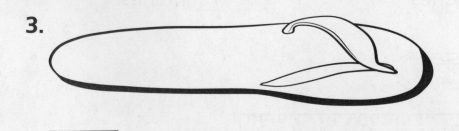

_____

Measure Length with Feet and Yards

Name _____

Review

You can use footprints to measure length in feet and yards. A footprint is 1 foot long. 3 footprints is 1 yard long.

What is the length of the surfboard?

Place footprints end-to-end under the surfboard. Count the footprints.

The surfboard is 6 feet long, or 2 yards long.

What is the length of the object in feet and yards?

1.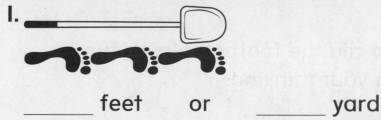

   _____ feet   or   _____ yard

2.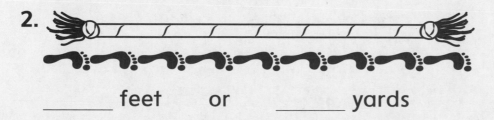

   _____ feet   or   _____ yards

Measure Length with Feet and Yards

Name _____

Which run shows how far the football player ran?

Run 1	Run 2	Run 3
1 yard	1 foot	2 yards

1.

Run _____

2.

Run _____

3.

Run _____

4. How many feet did the football player run on Run 3? Explain your thinking.

Compare Lengths Using Customary Units

Name _____

Review

You can use jumps along a ruler to compare length and write a subtraction equation.

Compare the length of the toy to the block.

The airplane is 5 inches longer than the block.

$7 - 2 = 5$

I. What equation can you write to compare the lengths? Fill in the numbers. Use the jumps to help.

_____ − _____ = _____

The pencil is _____ inches _____ than the pin.

Compare Lengths Using Customary Units

Name _____

Measure the objects in inches. Write the names of the objects in order from shortest to longest.

1. Spool of thread _____ inches

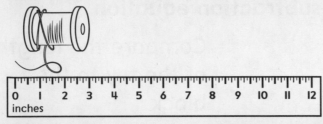

2. Toothbrush _____ inches

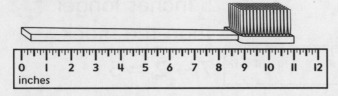

3. Dog treat _____ inches

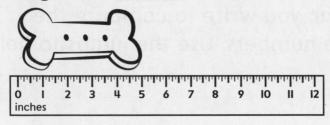

Order: _____

4. Compare the lengths. Write the names of the objects in order from longest to shortest.

   • Myra's bed is 7 feet long.

   • Paulette's garden is 14 feet long.

   • Rico's ladder is 17 feet long.

_____

Relate Inches, Feet, and Yards

Name _____

Review

You can relate inches, feet, and yards.

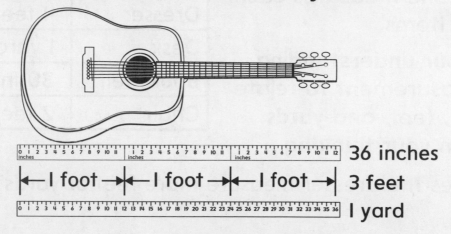

36 inches

|← I foot →|← I foot →|← I foot →| 3 feet

I yard

Inches are smaller than feet. It takes more inches than feet to measure the guitar.

Feet are smaller than yards. It takes more feet than yards to measure the guitar.

---

1. What is the height of the classroom door in yards?

   _____ yards

   Will it take more feet or yards to measure it?

   Circle the name of the unit.

   feet   yards

2. Will it take less inches or feet to measure the length of a car? Explain your thinking.

Relate Inches, Feet, and Yards

Name _____

Derinda wants to move things around in her room. She measures each of the items.

Use your understanding of measurement to relate inches, feet, and yards. Explain your thinking.

Item	Length
Bed	75 inches
Dresser	4 feet
Desk	1 yard
Bookshelf	30 inches
Chair	2 feet

1. Does the dresser measure more feet or yards?

2. Does the bed measure fewer inches or feet?

3. Does the desk measure more inches or yards?

4. Will the desk or chair take up more space? Explain.

Estimate Length Using Customary Units

Name _____

Review

You can use cubes to help you estimate length in inches.

Each cube is about 1 inch.

1 2 3 4 5 6

The wallet is about the same length as 6 cubes.

The wallet is about 6 inches long.

How long is the object? Estimate the length.

1.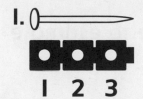

1 2 3

_____

2.

_ _ _ _ _

_____

Estimate Length Using Customary Units

Name _____

About how long is your desk?

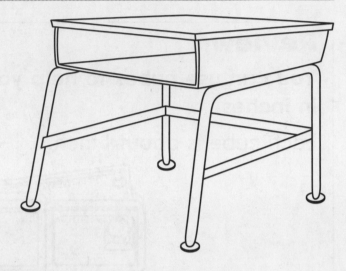

1. What are some objects you could use to estimate the length of your desk in inches?

2. What is a good estimate for the length of your desk in inches? Explain your thinking.

3. What are some objects you could use to estimate the length of your desk in feet?

4. What is a good estimate for the length of your desk in feet? Explain your thinking.

Measure Length with Centimeters and Meters

Name _____

Review

You can use grid paper to measure length in centimeters.

Each grid is I centimeter long.

What is the length of the ladybug?

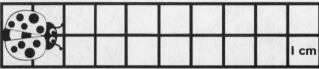

Place the ladybug on grid paper. Count the squares.

The ladybug is 2 centimeters long.

What is the length of the object in centimeters?

1.

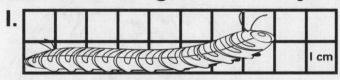

_____ centimeters

2.

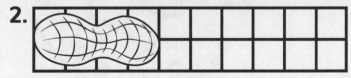

_____ centimeters

3. Sia wants to measure the length of her bedroom wall. Should she use a centimeter ruler or a meterstick? Explain.

Measure Length with Centimeters and Meters

Name _____

There are 5 coins in a collection. Which coin has the width shown? Use a centimeter ruler to measure from side to side.

nickel

penny

dime

quarter

half-dollar

1. 5 centimeters _____

2. 2 centimeters _____

3. 6 centimeters _____

4. 4 centimeters _____

5. 3 centimeters _____

Compare Lengths Using Metric Units

Name _____

Review

You can use jumps along a ruler to compare length and write a subtraction equation.

Compare the length of the egg to the orange slice.

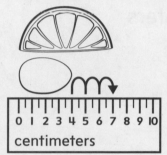

The egg is 3 centimeters shorter.

An equation is $7 - 4 = 3$.

---

1. What equation can you write to compare the lengths? Fill in the numbers.

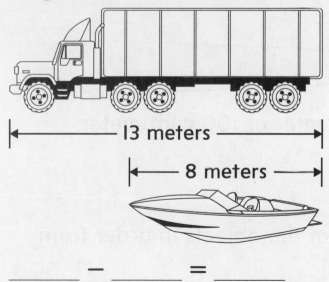

13 meters

8 meters

_____ − _____ = _____

The boat is _____ meters _____ than the truck.

Compare Lengths Using Metric Units

Name _____

Compare the length of the skateboard to the object.

skateboard =
70 centimeters

1. 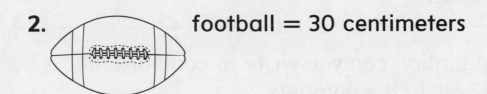   volleyball = 24 centimeters

2. football = 30 centimeters

3.

baseball bat = 1 meter or 100 centimeters

4. Write the names of the objects in order from shortest to longest.

Relate Centimeters and Meters

Name _____

Review

You can relate centimeters and meters.

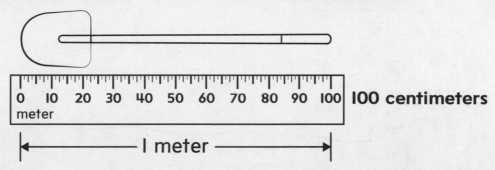

100 centimeters

meter

Meters are larger than centimeters. It takes fewer meters than centimeters to measure the shovel.

---

1. What is the length of your desk in centimeters?

_____ _____

2. Will it take fewer centimeters or meters to measure the desk? Circle the name of the unit.

centimeters   meters

3. Brent and Silvia want to measure Brent's baseball glove. Brent thinks there will be more meters. Silvia thinks there will be more centimeters. How do you respond to them?

Relate Centimeters and Meters

Name _____

The ruler represents I meter.

I meter = 100 centimeters

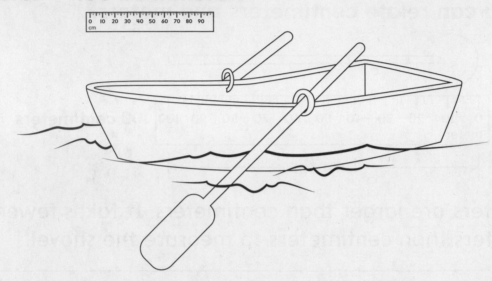

1. About how many meters will you need to measure the boat?

2. About how many centimeters will you need to measure the boat?

3. Do you need more centimeters or meters to measure the boat? Explain your thinking.

Estimate Length Using Metric Units

Name _____

Review

You can use fingers to help you estimate length in centimeters.

A fingernail is about I centimeter wide.

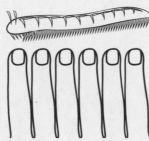

There are 6 fingernails.

The centipede is about 6 centimeters long.

I. How long is the object? Estimate the length.

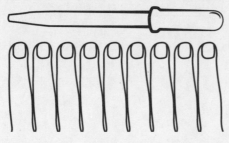

about _____ centimeters

2. Which unit would you use to measure the length of a school bus? Circle the correct answer.

centimeter   meter

Estimate Length Using Metric Units

Name _____

Name an object found in a house that is close to the length.

1. 5 centimeters long

2. 3 meters long

3. 30 centimeters long

4. 2 meters tall

5. 1 meter wide

6. 15 centimeters long

Solve Problems Involving Length

Name _____

Review

You can use base-ten blocks to solve problems involving length.

Ina has 15 feet of iron-on tape and 19 feet of stick-on tape.

How many feet of tape does Ina have in all?

Use base-ten blocks to add 15 + 19.

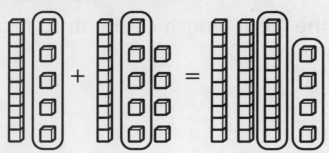

15 + 19 = **34**

Ina has **34** feet of tape.

---

1. Solve. Draw base-ten blocks to show your thinking.

Sid has 17 meters of yellow rope and 14 meters of blue rope. How many meters of rope does he have?

_____

Solve Problems Involving Length

Name

A designer takes some measurements in the room he is going to decorate.

Item	Length
sofa	74 inches
coffee table	38 inches
blue area rug	16 feet
green area rug	28 feet

Use the information in the table. How can you solve the problem? Solve. Explain your thinking.

1. What is the total length of the area rugs?

2. An end table is 25 inches long. How much longer is the coffee table than the end table?

3. Write another subtraction problem using 2 of the items from the table. Solve and explain your work.

Solve More Problems Involving Length

Name _____

Review

You can use part-part-whole mats to solve problems involving length.

A water snake is 71 centimeters long.

A garter snake is 55 centimeters long.

How much longer is the water snake than the garter snake?

Write a subtraction equation.	Write an addition equation.
$71 - 55 = ?$	$55 + ? = 71$

Part	Part
55	16
Whole	
71	

The water snake is 16 centimeters longer.

---

1. Solve the problem. Use a part-part-whole mat.

An electrician has 42 feet of wire. He uses 34 feet of wire. How much wire is left?

Part	Part
34	___
Whole	
42	

_____ _____

Solve More Problems Involving Length

Name _____

A builder is measuring the lengths of some of his supplies.

Supply	Length
Carpet	30 yards
Fence	29 feet
Copper wire	15 inches
Pipe	45 inches

How can you find the differences in lengths? Use the information in the table. Solve.

1. The builder uses 12 yards of carpet. How much is left? _____ _____

2. The builder buys 60 feet of fence. How much fence is there in all? _____ _____

3. Write another subtraction problem using 2 of the items from the table. Explain how to solve.

Understand the Values of Coins

Name _____

Review

You can find the value of a group of the same type of coin by skip counting.

Martina has 7 nickels. How many cents does she have?

 I nickel = 5¢

Use a number line to skip count by 5s. Make 7 jumps.

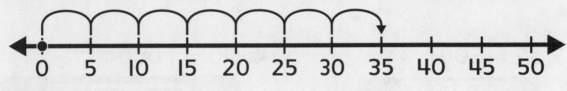

Martina has 35¢.

How many cents? Use the number line to help.

1. 13 pennies _____ ¢

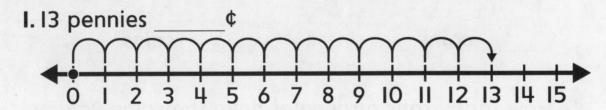

2. 12 dimes _____ ¢

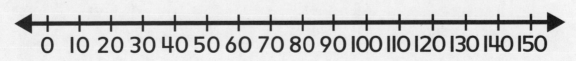

3. 9 nickels _____ ¢

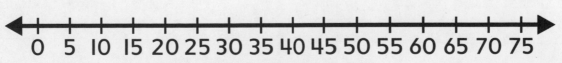

Understand the Values of Coins

Name _____

Banks and stores keep coins in rolls.

How many coins are in the roll?

1.

50¢ PENNIES 50¢

_____ coins

2. $2 = 200¢

$2 NICKLES $2

_____ coins

3. $5 = 500¢

$5 DIMES $5

_____ coins

4. $10 = 1,000¢

$10 QUARTERS $10

_____ coins

5. How many rolls of pennies have the same value as a roll of nickels? Explain your thinking.

Solve Money Problems Involving Coins

Name _____

Review

You can find the total value of coins by adding.

Adam has 1 quarter, 3 dimes, 2 nickels, and 3 pennies. How much money does he have?

Write the values in each box. Then add the values.

Quarters	Dimes	Nickels	Pennies
25¢	10¢ 10¢ 10¢	5¢ 5¢	1¢ 1¢ 1¢
25¢	**30¢**	**10¢**	**3¢**

Quarters and dimes: **25¢ + 30¢** = 55¢

Quarters, dimes, and nickels: 55¢ + **10¢** = 65¢

Quarters, dimes, nickels, and pennies: 65¢ + **3¢** = 68¢

So, Adam has 68¢.

What is the total value of the coins?

1. 2 quarters, 2 dimes, and 4 nickels

Quarters	Dimes	Nickels	Pennies
50¢	20¢	20¢	

total value: _____¢

2. 1 quarter, 2 dimes, 1 nickel, 4 pennies

Quarters	Dimes	Nickels	Pennies

total value: _____¢

Solve Money Problems Involving Coins

Name _____

Four students buy items from the school store.

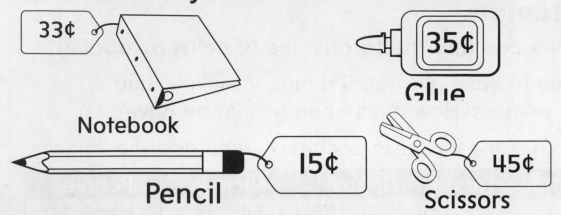

33¢ Notebook

35¢ Glue

15¢ Pencil

45¢ Scissors

What item does the student buy?

1.

   Curtis bought _____.

2.

   Javier bought _____.

3.

   Lucy bought a _____.

4.

   Maya bought a _____.

Solve Money Problems Involving Dollar Bills and Coins

Name _____

Review

You can add to find the total of like bills.

Van has three $20 bills, two $10 bills, and one $5 bill. How much money does he have?

Write the values in each box. Then add the values.

$20 bills	$10 bills	$5 bills	$1 bills
$20 $20 $20	$10 $10	$5	
$60	**$20**	**$5**	

20- and 10-dollar bills: **$60** + **$20** = $80

20-, 10-, and 5-dollar bills: $80 + **$5** = $85

So, Van has $85.

What is the total value of the bills?

1. four $10 bills, three $5 bills, two $1 bills

$20 bills	$10 bills	$5 bills	$1 bills
	$40	$15	$2

total value: $_____

2. two $20 bills, three $5 bills, four $1 bills

$20 bills	$10 bills	$5 bills	$1 bills

total value: $_____

Solve Money Problems Involving Dollar Bills and Coins

Name _____

What is the total value of the bills and coins?

1.

$ _____

2.

$ _____

3.

$ _____

4.

$ _____

Tell Time to the Nearest Five Minutes

Name _____

Review

You can tell time to the nearest five minutes.

What time is shown on the analog clock?

The hour hand points between 9 and 10. The hour is 9.

The minute hand points at 7.

Skip count by 5s 7 times:

5, 10, 15, 20, 25, 30, 35. The minute is 35.

The time shown on the analog clock is 9:35.

What time does the analog clock show?

1.

The hour hand points between _____ and _____. The hour is _____.

The minute hand points to _____. The minute is _____.

The time is _____ : _____.

2.

3.

_____ : _____   _____ : _____

Tell Time to the Nearest Five Minutes

Name _____

Can you answer the riddle? Write the letter that goes with the time shown.

H	O	T
quarter to 3:00	quarter past 3:00	

R	U	Y
	half past 4:00	

What time is the Mr. Molar's appointment with the dentist?

4:25	3:15	3:15	4:25	2:45

2:45	4:30	3:45	4:25	3:30

Be Precise When Telling Time

Name _____

Review

You can use a.m. and p.m. to describe times.

Use **a.m.** to represent the time between midnight and noon.

Use **p.m.** to represent the time from noon to midnight.

When might you eat lunch?	When might you go to bed?
You might eat lunch in the morning, at 11:45 **a.m.**	You might go to bed at nighttime, at 9:30 **p.m.**

What time of day does the event take place? Circle the correct time of day. Then write a.m. or p.m.

1. Volleyball practice takes place in the early morning/afternoon. Kendra might have volleyball practice at 4:15 _____

2. Rashaun just got a telescope for a gift. He plans to look at the stars in the afternoon/nighttime. Rashaun might look at the stars at 10:45 _____

3. The choir practices just before lunch, in the morning/nighttime. The choir might practice at 11:30 _____

Be Precise When Telling Time

Name _____

Give a possible time for the activity. Write a.m. or p.m. Draw a dot to show the time on the number line. Then draw a line from the activity to its matching time.

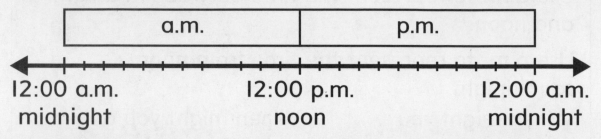

12:00 a.m.
midnight

12:00 p.m.
noon

12:00 a.m.
midnight

I. A tornado watch causes a delay for the fireworks show.

2. The Smith family runs outside to look at a rainbow in the sky.

_____

3. The baker drives to work through light fog.

4. Josh and Denni view an eclipse of the moon.

_____

5. Aaron and Selena view an eclipse of the sun.

6. The newspaper carrier walks through grass covered with dew.

_____

Use Mental Math to Add 10 or 100

Name _____

Review

You can add 10 or 100 using a place value chart.

234 + 10 = ?

hundreds	tens	ones
2	3	4

596 + 10 = ?

hundreds	tens	ones
5	9 + 1 = 10	6

To add 10, increase the tens digit by 1.

hundreds	tens	ones
2	3 + 1 = 4	4

10 tens is the same as 1 hundred and 0 tens.

hundreds	tens	ones
5 + 1 = 6	0	6

234 + 10 = 244

596 + 10 = 606

What is the sum? Fill in the chart to show your thinking.

1. 175 + 10 = _____

hundreds	tens	ones
_	_ + 1	_

2. 348 + 100 = _____

hundreds	tens	ones
_ + 1	_	_

3. 792 + 10 = _____

hundreds	tens	ones

4. 510 + 100 = _____

hundreds	tens	ones

What is the sum?

5. 687 + 100 = _____

6. 291 + 10 = _____

Use Mental Math to Add 10 or 100

Name _____

Sara plays a game. She starts with 50 points.
For each item she finds, she gets these points.

Bug = 1 point	Cat = 10 points	Star = 100 points

When she gets a total of 500 points, she earns
a trophy.

How many total points does Sara have after the
level? Circle the level number when she earns
a trophy.

1. Level 1: Sara finds 🐱 and ⭐.

   50 + _____ + _____ = _____

2. Level 2: Sara finds 🐱, 🐱, and 🐞.

3. Level 3: Sara finds ⭐, 🐱, and ⭐.

4. Level 4: Sara finds 🐱 and ⭐.

5. Level 5: Sara finds 🐞, ⭐, and 🐞.

Represent Addition with 3-Digit Numbers

Name _____

Review

You can use a place-value chart to help you add 3-digit numbers.

$342 + 136 = ?$

hundreds	tens	ones
3	4	2
1	3	6
4	7	8

Write the digits of the addends in the chart.

Add down the columns to find the number of hundreds, tens, and ones in the sum.

$342 + 136 = 478$

What is the sum? Use the chart to show your thinking.

1. $122 + 345 =$ _____

hundreds	tens	ones
1	2	2
3	4	5

2. $617 + 251 =$ _____

hundreds	tens	ones

What is the sum?

3. $450 + 332 =$ _____   **4.** $376 + 610 =$ _____

Represent Addition with 3-Digit Numbers

Name _____

**Jaylen wins 800 prize tickets at a fun center.
The prizes below cost the number of tickets shown.**

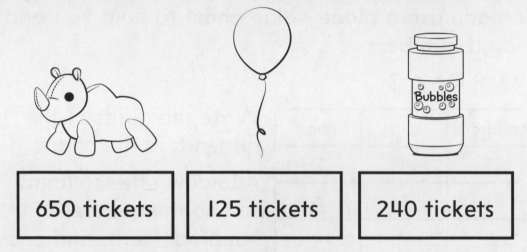

| 650 tickets | 125 tickets | 240 tickets |

1. Does Jaylen have enough prize tickets for the stuffed animal and the balloon? Explain.

2. Does Jaylen have enough prize tickets for 2 balloons and 2 bubbles? Explain.

3. If Jaylen won 100 more tickets, would she have enough prize tickets for the stuffed animal and the bubbles? Explain.

Represent Addition with 3-Digit Numbers with Regrouping

Name _____

Review

A place-value chart can help you add 3-digit numbers.

$394 + 251 = ?$

Add down the columns to find the number of hundreds, tens, and ones in the sum.

hundreds	tens	ones
3	9	4
2	5	1
5	(14)	5

14 tens is the same as 140. Regroup 14 tens as 4 tens and 1 hundred.

hundreds	tens	ones
5 (+1)	(4)	5

$394 + 251 = 645$

What is the sum? Fill in the table to help you regroup.

1. $267 + 318 =$ _____

hundreds	tens	ones
2	6	7
3	1	8
5	7	(15)
__	__	__

2. $394 + 261 =$ _____

hundreds	tens	ones
3	9	4
2	6	1
__	(__)	__
__	__	__

Represent Addition with 3-Digit Numbers with Regrouping

Name _____

Will you have to regroup the ones, the tens, or both to find the sum? Circle your answer. Then, find the sum.

1. $657 + 251 = ?$ ones   tens   both

    _____

2. $218 + 553 = ?$ ones   tens   both

    _____

3. $338 + 285 = ?$ ones   tens   both

    _____

4. $335 + 265 = ?$ ones   tens   both

    _____

What is the sum?

5. $245 + 329 + 130 =$ _____

6. $173 + 269 + 518 =$ _____

Decompose Addends to Add 3-Digit Numbers

Name _____

Review

You can count by place value to add 3-digit numbers.

268 + 453 = ?

Start at 268 and count on by 100s. 453 has four 100s:

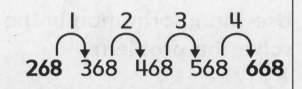

From **668**, count on by 10s. 453 has five 10s:

From **718**, count on by 1s. 453 has three 1s:

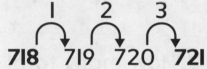

268 + 453 = **721**

What is the sum? Count by place value.

1. 164 + 213 = _____

   Count on by 100s from _____: _____ _____

   Count on by 10s from _____: _____

   Count on by 1s from _____: _____ _____ _____

2. 359 + 234 = _____

   Count on by 100s from _____: _____ _____

   Count on by 10s from _____: _____ _____ _____

   Count on by 1s from _____: _____ _____ _____ _____

Differentiation Resource Book

Decompose Addends to Add 3-Digit Numbers

Name

The table shows the number of customers at an ice cream shop over a 3-day holiday weekend.

Day	Customers
Friday	255
Saturday	328
Sunday	84

Use the information in the table. Decompose to solve the problem.

1. How many total customers were served on Friday and Sunday?

   _____ customers

2. How many total customers were served on Saturday and Sunday?

   _____ customers

3. How many customers were served in all during the 3-day weekend? Explain how you found your answer.

   _____ customers

Decompose One Addend to Add 3-Digit Numbers

Name _____

Review

You can decompose one addend and add by place value.

$342 + 234 = ?$

Decompose the second addend. Use a place-value chart to show the decomposed addend.

$234 =$

hundreds	tens	ones
2	3	4

Add 2 hundreds to 342. $\longrightarrow 342 + 100 + 100 = 542$

Add 3 tens to the result. $\longrightarrow 542 + 10 + 10 + 10 = 572$

Then add 4 ones. $\longrightarrow 572 + 1 + 1 + 1 + 1 = 576$

$342 + 234 = 576$

What is the sum? Decompose one addend and show the result in a place-value chart. Then add by place value.

1. $135 + 422 =$ _____

hundreds	tens	ones
4	2	2

2. $472 + 317 =$ _____

hundreds	tens	ones

Decompose One Addend to Add 3-Digit Numbers

Name _____

One of the addends in each sum has been decomposed. Draw a line from the sum to its decomposition. Then use the letters to solve the riddle.

1. $257 + 367$

2. $257 + 355$

3. $355 + 367$

4. $367 + 165$

5. $165 + 355$

$165 + 300 + 50 + 5$
N Sum: _____

$257 + 300 + 50 + 5$
T Sum: _____

$367 + 200 + 50 + 7$
E Sum: _____

$367 + 100 + 60 + 5$
O Sum: _____

$355 + 300 + 60 + 7$
S Sum: _____

Write the sums from least to greatest to solve the riddle.

Riddle: Why did the music teacher bring a ladder to class?

Answer: To reach the high _____!

Adjust Addends to Add 3-Digit Numbers

Name _____

Review

You can adjust addends and use a number line to add.

204 + 318 = ?

| −4 | | +4 |

200 + 322 = ?

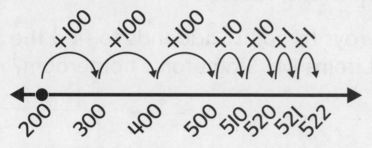

204 + 318 = **522**

What is the sum? Adjust one addend. Use a number line to show how to add the adjusted numbers.

1. 297 + 123 = ___

| +3 | | −3 |

300 + 120 = ___

⟷

2. 495 + 206 = ___

| | | |

500 + ___ = ___

⟷

Adjust Addends to Add 3-Digit Numbers

Name _____

The table shows the number of items each homeroom collects during a charity food drive. The homeroom that collects the most food items wins a prize.

Homeroom	Number of Items	
	Day 1	Day 2
Ms. Patterson	151	147
Mrs. Hillshire	198	142
Mr. Singleton	166	141

Use the information in the table.

1. What are two ways to adjust addends to find the number of food items Mr. Singleton's homeroom?

2. Which homeroom won the contest? Circle the name. How do you know?

3. The grade wins a prize if they collect more than 500 items in a day. On which days did the grade win a prize? Explain your thinking.

Explain Addition Strategies

Name _____

Review

You can choose the best strategy to add 3-digit numbers.

Adjust the Addends	Place-Value Table

Adjust the Addends

$321 + 267 =$

$321 + 267 = ?$

⬚ −1   ⬚ +1

$320 + 268 = 588$

Place-Value Table

$395 + 257 =$

hundreds	tens	ones
5 + 1	4 + 1	2

$395 + 257 = 652$

How can you find the sum? Circle the best strategy. Then find the sum.

1. $358 + 475 =$ _____

   adjust the addends

   use a place-value chart

2. $241 + 326 =$ _____

   adjust the addends

   use a place-value chart

3. $469 + 126 =$ _____

   adjust the addends

   use a place-value chart

Explain Addition Strategies

Name _____

Choose a strategy to solve the problem. Explain your thinking.

1. Janie took 478 steps walking to Maya's house. Then she took 344 steps to the park. How many steps did she take in all?

   _____ steps

2. Rodrigo found 215 gold coins on the first level of a video game. He found 183 gold coins on the second level. How many gold coins did he find in all?

   _____ gold coins

3. Pam has 197 stickers in one book and 148 stickers in another book. How many stickers does she have in all?

   _____ stickers

Use Mental Math to Subtract 10 and 100

Name _____

Review

You can subtract 10 or 100 using a place value chart.

$537 - 100 = ?$ $614 - 10 = ?$

$437 - 100 = ?$ $604 - 10 = ?$

The hundreds digit goes down by 1.

The tens digit goes down by 1.

hundreds	tens	ones
$5 - 1 = 4$	3	7

$537 - 100 = 437$

hundreds	tens	ones
6	$1 - 1 = 0$	4

$614 - 10 = 604$

hundreds	tens	ones
$4 - 1 = 3$	3	7

$437 - 100 = 337$

hundreds	tens	ones
$6 - 1 = 5$	$10 - 1 = 9$	4

$604 - 10 = 594$

If there are 0 tens, the tens digit changes to 9 and the hundreds digit goes down by 1.

What is the difference? Fill in the chart.

1. $173 - 10 =$ _____

hundreds	tens	ones
1	$7 - 1 = \_$	3

2. $271 - 100 =$ _____

hundreds	tens	ones
$\_ - 1 = \_$	7	1

Use Mental Math to Subtract 10 and 100

Name _____

Some friends are shopping at the bookstore. Write the correct name beneath the piggy bank to show how many pennies the student has left after shopping.

225 pennies   325 pennies   335 pennies   385 pennies

_____   _____   _____   _____

- Nikki has 435 pennies. She spends 100 pennies on a folder and 10 pennies on a pencil.

- Raul has 425 pennies. He spends 100 pennies on paper and 100 pennies on a colored pencil.

- Raj has 415 pennies. He spends 10 pennies on a crayon, 10 pennies on a pencil, and 10 pennies on a stamp.

- Amy has 455 pennies. She spends 100 pennies on stickers, 10 pennies on an eraser, and 10 pennies on a bookmark.

Represent Subtraction with 3-Digit Numbers

Name _____

Review

You can use a place-value chart to help you subtract 3-digit numbers.

$584 - 244 = ?$

Write the digits of the numbers in the chart.

hundreds	tens	ones
$5 - 2 = 3$	$8 - 4 = 4$	$4 - 4 = 0$

Subtract the hundreds.
Subtract the tens.
Subtract the ones.

$584 - 244 = 340$

Use the chart to find the difference.

1. $347 - 123 =$ _____

hundreds	tens	ones
$3 - 1 = \_$	$4 - 2 = \_$	$7 - 3 = \_$

2. $785 - 314 =$ _____

hundreds	tens	ones
$7 - 3 = \_$	$8 - 1 = \_$	$5 - 4 = \_$

3. $566 - 235 =$ _____

hundreds	tens	ones
$\_ - \_ = \_$	$\_ - \_ = \_$	$\_ - \_ = \_$

4. $449 - 243 =$ _____

hundreds	tens	ones
$\_ - \_ = \_$	$\_ - \_ = \_$	$\_ - \_ = \_$

Represent Subtraction with 3-Digit Numbers

Name _____

Write a subtraction word problem for the representation. Then solve the problem.

1.

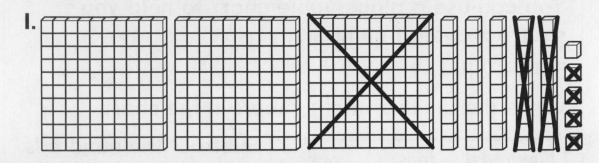

2.

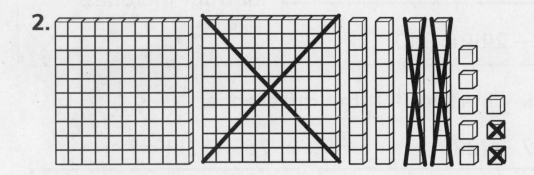

3.

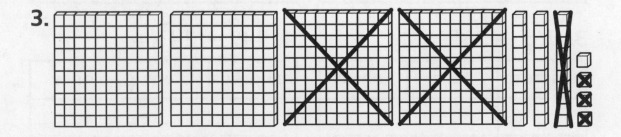

Decompose One 3-Digit Number to Count Back

Name _____

Review

You can count back by place value to subtract 3-digit numbers.

354 − 235 = ?

Start at 354. Count back by 100s.

There are 2 100s in 235.   354   254   **154**

Start at 154. Count back by 10s.

There are 3 10s in 235.   154   144   134   **124**

Start at 124. Count back by 1s.

There are 5 1s in 235.   124   123   122   121   120   **119**

354 − 235 = 119

---

1. What is the difference? Count back by place value.

803 − 327 = _____

Count back by 100s from 803: _____, 603, _____

Count back by 10s from _____: 493, _____

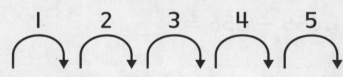

Count back by 1s from _____: _____, 481, _____,

_____, 478, _____, _____

Decompose One 3-Digit Number to Count Back

Name _____

Find the difference. Then write the letter from the table that matches the difference.

Difference	150	299	302	319	455	457	625	776
Letter	A	C	E	I	K	R	S	T

1. $438 - 139 =$ _____   letter: _____

2. $713 - 256 =$ _____   letter: _____

3. $515 - 365 =$ _____   letter: _____

4. $655 - 356 =$ _____   letter: _____

5. $967 - 512 =$ _____   letter: _____

Write the letters in order to solve the riddle.

Riddle: Why shouldn't you tell a joke to an egg?

Answer: It might _____ up!

Counting On to Subtract 3-Digit Numbers

Name _____

Review

You can write a related addition equation to subtract.

$356 - 231 = ?$

Write a related addition equation. $231 + ? = 356$

Show 231. Then add to make 356.

hundreds	tens	ones
□ □ □	‖‖	• ⋮

Find the sum added to 231: $100 + 20 + 5 = 125$.

Since $231 + 125 = 356$, $356 - 231 = 125$.

What is the difference? Write a related addition equation. Use ? for the unknown. Show how to count on.

1. $547 - 322 =$ _____

Equation: $322 + ? = 547$

hundreds	tens	ones
□ □ □	‖	⦂

2. $320 - 114 =$ _____

Equation: _____

hundreds	tens	ones

Counting On to Subtract 3-Digit Numbers

Name _____

Write a subtraction equation that is modeled by the number line. Then write the related addition equation.

1.

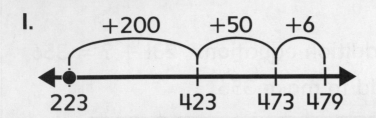

+200   +50   +6

223   423   473 479

2.

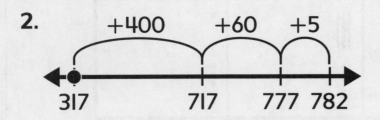

+400   +60   +5

317   717   777 782

3.

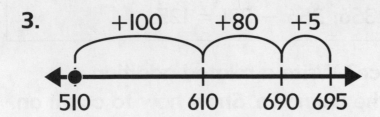

+100   +80   +5

510   610   690 695

4. Explain how you can count on to subtract 3-digit numbers.

Regroup Tens

Name _____

Review

Sometimes you need to regroup a ten when subtracting.

$653 - 128 = ?$

Use base-ten shorthand to show 653.

Regroup I ten into 10 ones.

Then subtract.

$653 - 128 = 525$

What is the difference? Show how to regroup tens by drawing base-ten shorthand.

1. $564 - 258 =$ _____

2. $638 - 419 =$ _____

Regroup Tens

Name _____

The toys cost the number of pennies shown.

Bubbles   Toy Plane   Bucket

| 125 pennies | 375 pennies | 175 pennies |

1. Caleb has 454 pennies. If he buys bubbles, how many pennies will he have left?

2. Julia has 691 pennies. If she buys a bucket, how many pennies will she have left?

3. Lance has 480 pennies. He buys a toy plane. Can he also buy bubbles? Explain.

4. Do you have to regroup tens to solve each of these problems? Explain why or why not.

Regroup Tens and Hundreds

Name _____

Review

Sometimes you need to regroup tens and hundreds when subtracting.

$345 - 168 = ?$

Use base-ten shorthand to show 345.

Regroup 1 hundred into 10 tens.

Regroup 1 ten into 10 ones.

Then subtract.

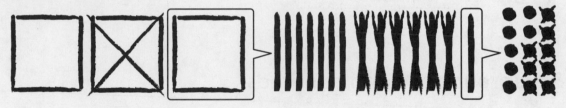

$345 - 168 = 177$

What is the difference? Show how to regroup tens and hundreds by drawing base-ten shorthand.

1. $455 - 170 =$ _____

2. $546 - 159 =$ _____

Regroup Tens and Hundreds

Name _____

Write a 3-digit number to be subtracted so that the given digits have to be regrouped. Then find the difference.

1. Regroup tens only:

   553 − _____ = _____

2. Regroup hundreds only:

   736 − _____ = _____

3. Regroup tens and hundreds:

   437 − _____ = _____

4. Regroup tens and hundreds:

   631 − _____ = _____

5. Describe when you have to regroup tens and hundreds to subtract 3-digit numbers.

Adjust Numbers to Subtract 3-Digit Numbers

Name _____

Review

You can adjust numbers to make them easier to subtract.

$467 - 198 = ?$

$\boxed{+2}$ $\boxed{+2}$

$469 - 200 = ?$

$467 - 198 = 269$

hundreds	tens	ones
□ □ ⊠ ⊠	|||||	⦂⦂⦂

What is the difference? Adjust one of the numbers to subtract. Show the subtraction with base-ten shorthand.

1. $279 - 147 =$ ____

$\boxed{+3}$ $\boxed{+3}$

$282 - 150 =$ ____

hundreds	tens	ones
□ ⊠	|||| XXXXX	:

2. $354 - 205 =$ ____

$\boxed{}$ $\boxed{}$

____ $- 200 =$ ____

hundreds	tens	ones

Adjust Numbers to Subtract 3-Digit Numbers

Name _____

Find the difference. Then write the letter from the table that matches the difference.

Difference	118	394	268	375	272	175	244	386
Letter	W	T	V	A	C	B	S	E

1. 315 − 197 = _____   letter: _____

2. 522 − 147 = _____   letter: _____

3. 403 − 135 = _____   letter: _____

4. 796 − 410 = _____   letter: _____

5. 647 − 403 = _____   letter: _____

Write the letters in order to solve the riddle.

Riddle: How do you know the ocean is so friendly?

Answer: It always _____ at you!

Explain Subtraction Strategies

Name _____

Review

You can use different strategies to subtract numbers.

$485 - 253 = ?$

Use base-ten shorthand to show 485.

Subtract the hundreds, tens, and ones.

$485 - 253 = 232$

What is the difference? Use the suggested strategy.

1. Write a related addition equation. Use ? for the unknown. Then count on to find the difference.

   $659 - 321 =$ _____

   Equation: _____

2. Decompose the number being subtracted. Then count back to subtract.

   $966 - 450 =$ _____

   $450 =$ _____ $+$ _____

Explain Subtraction Strategies

Name _____

Choose a strategy to solve the problem. Explain your thinking.

1. Maria reads to page 256 of her book. She starts on page 147. How many pages does she read?

2. A vending machine has 368 bouncy balls at the start of the week. By the end of the week, 104 balls are sold. How many bouncy balls are left?

3. The school cafeteria holds a total of 288 students. Currently, there are 145 students in the cafeteria. How many more students will fit?

Solve Problems Involving Addition and Subtraction

Name _____

Review

You can use addition and subtraction strategies to solve word problems.

James prints 250 flyers. He hands out 112 flyers. His sister hands out 105 flyers. How many flyers are left?

You can solve the problem with **addition** and **subtraction**.

$105 + 112 = 217$

$250 - 217 = 33$

There are 33 flyers left.

Or, you can solve the problem with **subtraction only**.

$250 - 112 = 138$

$138 - 105 = 33$

Solve the problem.

1. Trey's video game character has 255 coins. He spends 135 coins. Then he earns 202 coins. How many coins does the character have now?

   $255 - 135 =$ _____   _____ $+ 202 =$ _____

2. Molly has 395 marbles. Isabela has 122 more marbles than Molly. Collin has 147 fewer marbles than Isabela. How many marbles does Collin have?

Solve Problems Involving Addition and Subtraction

Name

Write a word problem that requires the operation or operations given. Then use any strategy to solve it.

I. Write a one-step word problem that uses addition only.

2. Write a one-step word problem that uses subtraction only.

3. Write a two-step word problem that uses both addition and subtraction.

Understand Picture Graphs

Name _____

> ## Review
> You can represent data in a picture graph.
>
> Shade one square for each tally in the tally chart.

How can you represent the data in a picture graph?

Favorite Flavor	
Flavor	**Tally**
Vanilla	HHT I
Chocolate	HHT II
Strawberry	IIII
Mint	I

Favorite Flavor							
Vanilla							
Chocolate							
Strawberry							
Mint							

Each square = 1 vote

Understand Picture Graphs

Name

A group of children were asked to vote for their favorite farm animal.

Favorite Farm Animal

Goat	🐐	🐐	🐐				
Cow	🐄	🐄	🐄	🐄	🐄	🐄	
Pig	🐖	🐖	🐖	🐖	🐖	🐖	🐖
Chicken	🐔	🐔	🐔	🐔	🐔		

Each picture = 1 vote

Explain why the sentence is *not* true.

1. Donna voted for rabbit.

2. Ellie, Kaylie, Josh, and Clayton said they all voted for goat.

3. There were 14 boys and 9 girls who voted. Every boy voted for either cow or chicken.

Understand Bar Graphs

Name _____

Review

You can **represent data in a bar graph.**

Shade each bar length to match the tally amount..

I. How can you represent the data in a bar graph?

Favorite Game	
Game	**Tally**
Board	IIII
Card	II
Dice	IIII I
Video	III

Favorite Game

Game							
Board							
Card							
Dice							
Video							

0 1 2 3 4 5 6

Number

Understand Bar Graphs

Name _____

The picture graph shows the number of carnival prizes that were won.

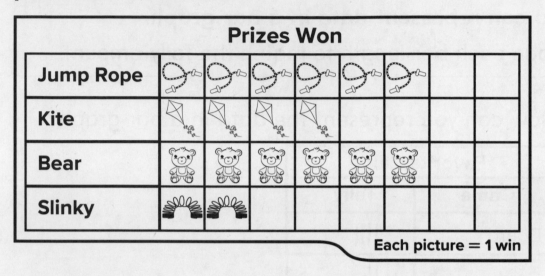

Prizes Won

Jump Rope	🪢	🪢	🪢	🪢	🪢	🪢		
Kite	🪁	🪁	🪁	🪁				
Bear	🧸	🧸	🧸	🧸	🧸	🧸		
Slinky	🌀	🌀						

Each picture = 1 win

I. Make a vertical bar graph of the data.

2. How many more kites need to be won so that the number won is the same as the prize with the most won.

Solve Problems Using Bar Graphs

Name

Review

You can use a bar graph to solve problems.

What school lunch was chosen the most?

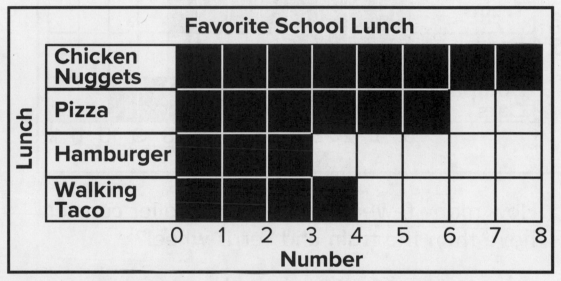

The bar for chicken nuggets is the longest.
So, it was chosen the most.

Use the data in the bar graph to solve. Explain.

1. What school lunch was chosen the least?

2. How many more students chose pizza than walking tacos?

Solve Problems Using Bar Graphs

Name _____

Use the bar graph to solve the problems.

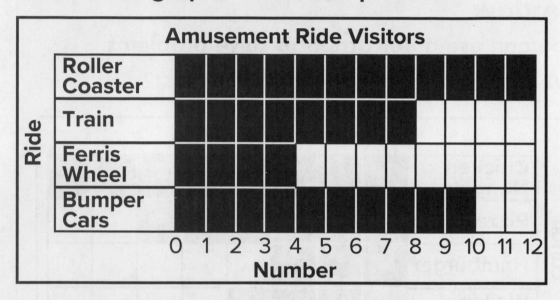

1. How many fewer visitors did the roller coaster have than the train and Ferris wheel?

2. How many more visitors did the Ferris wheel and bumpers cars have than the train?

3. How many fewer visitors did the train have than the roller coaster, Ferris wheel, and bumper cars?

4. Suppose there were 15 visitors to the roller coaster. How would you change the bar graph?

Collect Measurement Data

Name _____

Review

You can make a tally chart of the data.

○
47 inches   49 inches
48 inches   50 inches
51 inches   47 inches
47 inches   48 inches
○ 50 inches   48 inches
47 inches   49 inches
48 inches   48 inches
○ 48 inches   49 inches

Make one tally mark for each measure.

Heights of Students	
Height (inches)	**Tally**
47	IIII
48	HHT I
49	III
50	II
51	I

1. Mia measured the length of her hair ribbons. How can you make a tally chart to show the data?

○
16 inches   12 inches
12 inches   12 inches
18 inches   16 inches
20 inches   16 inches
○ 18 inches   18 inches
12 inches   18 inches
18 inches   18 inches

Length of Ribbons	
Length (inches)	**Tally**
12	
16	
18	
20	

Collect Measurement Data

Name _____

You can use the lengths of pencils to make a tally chart.

1. Measure the lengths of different pencils in centimeters. Make a list of the data.

2. Use your data to make a tally chart.

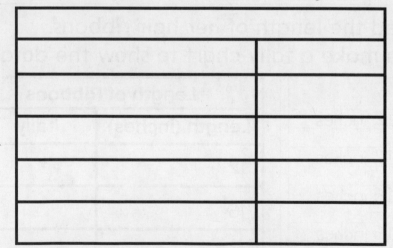

3. Write a question about the data in your tally chart. Then answer your question.

Understand Line Plots

Name _____

Review

You can use a line plot to answer questions.

What is the most common distance from school?

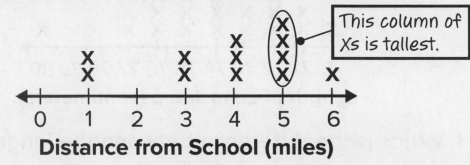

This column of Xs is tallest.

Distance from School (miles)

The most common distance from school is 5 miles.

Use the line plot about distance from school to answer the questions.

1. What is the longest distance recorded?

2. What distance do 3 students live from school?

3. How many distances were recorded?

4. What is the shortest distance recorded?

5. How many students live 2 miles away? Explain.

Understand Line Plots

Name _____

The physical education teacher records the lengths of the standing long jump of her students.

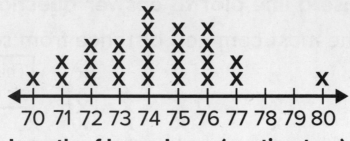

Length of Long Jump (centimeters)

1. What is the difference between the longest and shortest jumps recorded?

   _____ centimeters

2. How many students had a jump longer than the most common jump length?

   _____ students

3. Why do you think there is only 1 student that had a jump length of 80 centimeters?

4. How would the line plot change if a student who was absent jumps 68 centimeters? Explain.

Show Data on a Line Plot

Name _____

Review

You can use cubes to make a line plot.

Tia measured the lengths of stuffed animals in inches.

Place one cube above each measurement in the list.

16 inches	13 inches
10 inches	16 inches
14 inches	9 inches
13 inches	14 inches
16 inches	10 inches
9 inches	10 inches
8 inches	18 inches
13 inches	16 inches

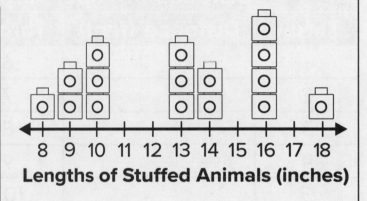

Lengths of Stuffed Animals (inches)

Answer the questions about Tia's line plot.

1. What is the title of the line plot?

2. Why does the number line begin at 8 and end at 18? Explain.

Show Data on a Line Plot

Name _____

Amir made a paper airplane. He threw the paper airplane 10 times.

1. What distances do you think Amir's paper airplane was thrown each time? Make a table of the data.

Distance of Paper Airplane Throw			
Throw	Distance (feet)	Throw	Distance (feet)
1		6	
2		7	
3		8	
4		9	
5		10	

2. Make a line plot of the data.

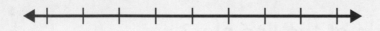

3. How can you be sure each throw is represented in the line plot?

Recognize 2-Dimensional Shapes by Their Attributes

Name _____

Review

You can recognize 2-dimensional shapes by their sides, angles, and vertices.

How many sides, angles, and vertices does the shape have?

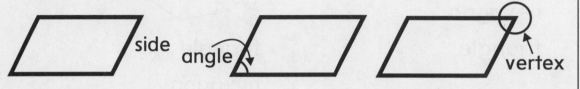

There are 4 sides, 4 angles, and 4 vertices.

How many sides, angles, and vertices does the shape have?

1.

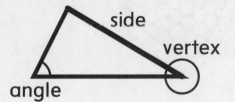

_____ sides

_____ angles

_____ vertices

2.

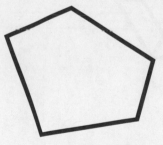

_____ sides

_____ angles

_____ vertices

Recognize 2-Dimensional Shapes by Their Attributes

Name _____

Circle the name of the shape of the object.

1.

quadrilateral

hexagon

triangle

2.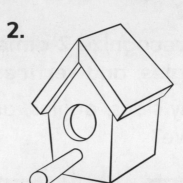

triangle

hexagon

pentagon

3.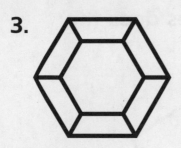

triangle

hexagon

pentagon

4.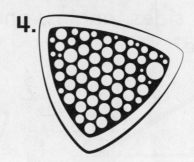

triangle

hexagon

quadrilateral

5. Find an object around your school or at home. What shape is the object? Explain how you know.

Draw 2-Dimensional Shapes from Their Attributes

Name _____

Review

You can draw a 2-dimensional shape given its attributes.

Draw a shape that has 3 sides, 3 angles, and all sides different lengths.

Draw a side.

Draw a second side that has a different length.

Draw a third side to make a triangle. Check that all sides have different lengths.

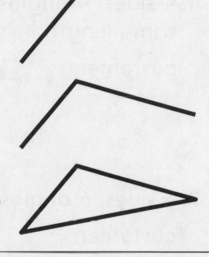

Draw the shape. Then write the name.

1. 4 sides, 4 angles, and all sides the same length

2. 5 sides, 5 angles, and all sides the same length

_____   _____

Draw 2-Dimensional Shapes from Their Attributes

Name _____

Joe is building some custom picture frames for friends. Draw the shape. Then, write the name of the friend that matches the set of attributes.

Customer	Shape
Cindy	square
Seth	hexagon
Mickie	rectangle

1. 4 sides, 4 angles, and opposite sides the same length

   customer: _____

2. 6 sides, 6 angles, and all sides the same length

   customer: _____

3. 4 sides, 4 angles, and all sides the same length

   customer: _____

Recognize 3-Dimensional Shapes by Their Attributes

Name _____

Review

You can recognize 3-dimensional shapes by their faces, edges, and vertices.

How many faces, edges, and vertices does this shape have? Name the shape.

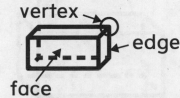

There are 6 rectangular with faces, 12 edges, and 8 vertices.

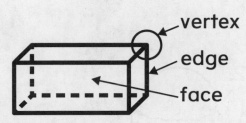

The shape is a rectangular prism.

How many faces, edges, and vertices does the shape have? Name the shape.

1.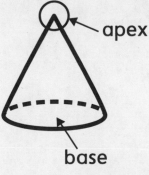

apex

base

_____ base

_____ edges

_____ apex

The shape is a _____.

2.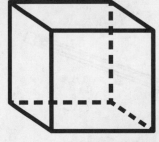

_____ square faces

_____ edges

_____ vertices

The shape is a _____.

Recognize 3-Dimensional Shapes by Their Attributes

Name _____

Some 3-dimensional shapes are made of more than one shape. What shapes make up the object? Explain.

1.

2.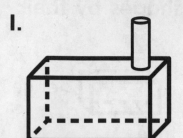

3.

Riddle: What is the coolest shape in town?

Answer: an ice _____!

Understand Equal Shares

Name _____

Review

You can partition shapes into equal shares.

How can you partition a circle into 4 equal shares?

Draw a line that partitions the circle into 2 equal shares or halves.

Draw another line that partitions the circle into 4 equal shares or fourths.

1. How can you partition the shape into 4 equal shares?

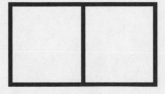

Draw a line to partition the rectangle into _____.

Then draw a line that creates _____.

Understand Equal Shares

Name _____

Partition the shape in two different ways. Draw to show your work.

1. 4 equal shares

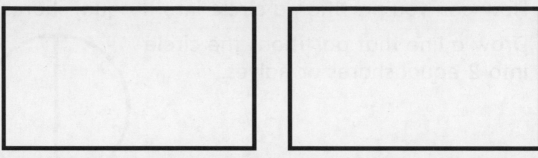

2. 4 equal shares

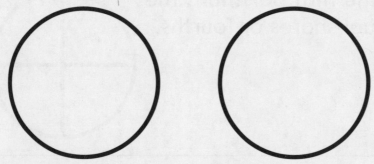

3. Randy partitions a square into 4 equal shares that are smaller squares. Olivia partitions the square into 4 equal shares that are triangles. They each draw 2 lines. Draw to show their partitions.

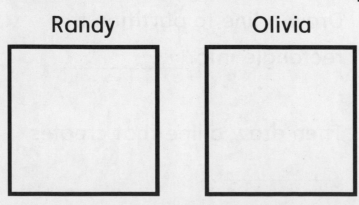

Relate Equal Shares

Name

Review

You can partition shapes in different ways.

There are different ways to show fourths.

How can you partition each shape into equal shares? Draw to show 2 different ways.

1. Partition the same circle into halves.

Draw a vertical line.   Draw a horizontal line.

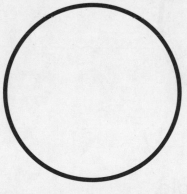 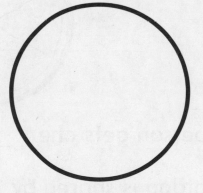

2. Partition the same rectangle into thirds.

Relate Equal Shares

Name _____

The food items will be shared equally. Draw to show how to partition the item. Then describe how much each person gets.

1. The pizza is shared by 4 people.

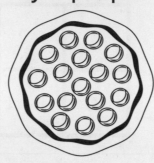

Each person gets one _____.

2. The pie is shared by 3 people.

Each person gets one _____.

3. The hotdog is shared by 2 people.

Each person gets one _____.

Partition a Rectangle into Rows and Columns

Name _____

Review

You can use rows and columns of squares to make a rectangle.

How many rows, columns, and squares is the rectangle partitioned into?

There are **2 rows**. Each row has 4 squares in it.

4 + 4 = 8 squares

There are **4 columns**. Each column has 2 squares in it.

2 + 2 + 2 + 2 = 8 squares

Use repeated addition to find the number of squares.

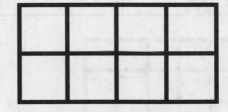

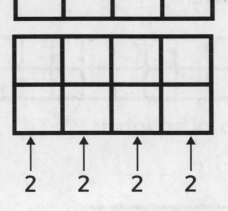

1. How many rows, columns, and total squares is the rectangle partitioned into?

Rows: _____

Columns: _____

Total squares: 3 + 3 = _____

Partition a Rectangle into Rows and Columns

Name _____

**Find the total number of squares in the rectangle.
Write the letter from the table that matches the total.**

Total	10	12	14	15	16	18	20	22
Letter	T	E	A	C	N	P	S	F

1.

total squares: _____

letter: _____

2.

total squares: _____

letter: _____

3.

total squares: _____

letter: _____

4.

total squares: _____

letter: _____

5.

total squares: _____

letter: _____

Write the letters in order to solve the riddle.

Riddle: Why didn't the quarter roll down the hill with the dime?

Answer: Because it had more _____!